# GAME THEORY

Other books from Automatic Press ♦ $\frac{V}{I}$P

**Formal Philosophy**
edited by Vincent F. Hendricks & John Symons
November 2005

**Thought$_2$Talk: A Crash Course in Reflection and Expression**
by Vincent F. Hendricks
September 2006

**Masses of Formal Philosophy**
edited by Vincent F. Hendricks & John Symons
October 2006

**Political Questions: 5 Questions for Political Philosophers**
edited by Morten Ebbe Juul Nielsen
December 2006

**Philosophy of Technology: 5 Questions**
edited by Jan-Kyrre Berg Olsen & Evan Selinger
February 2007

**Philosophy of Mathematics: 5 Questions**
edited by Vincent F. Hendricks & Hannes Leitgeb
August 2007

**Normative Ethics: 5 Questions**
edited by Jesper Ryberg & Thomas S. Petersen
August 2007

**Legal Philosophy: 5 Questions**
edited by Ian Farrell & Morten Ebbe Juul Nielsen
August 2007

**Philosophy of Physics: 5 Questions**
edited by Juan Ferret & John Symons
September 2007

**Probability and Statistics: 5 Questions**
edited by Alan Hájek & Vincent F. Hendricks
March 2008

# GAME THEORY
# 5 QUESTIONS

edited by

Vincent F. Hendricks

Pelle G. Hansen

Automatic Press ◆ $\frac{V}{I}$P

Automatic Press ♦ $\frac{V}{I}$P

Information on this title: www.gametheorists.com

© Automatic Press / VIP 2007

This publication is in copyright. Subject to statuary exception
and to the provisions of relevant collective licensing agreements,
no reproduction of any part may take place without
the written permission of the publisher.

First published 2007

Printed in the United States of America
and the United Kingdom

ISBN-10   87-991013-4-3   paperback

The publisher has no responsibilities for
the persistence or accuracy of URLs for external or
third party Internet Web sites referred to in this publication
and does not guarantee that any content on such
Web sites is, or will remain, accurate or appropriate.

Typeset in $\LaTeX 2_\varepsilon$
Cover photo and graphic design by Vincent F. Hendricks

# Contents

| | | |
|---|---|---|
| Preface | | iii |
| Acknowledgements | | v |
| 1 | Robert Aumann | 1 |
| 2 | Johan van Benthem | 9 |
| 3 | Cristina Bicchieri | 21 |
| 4 | Ken Binmore | 33 |
| 5 | Adam Brandenburger | 41 |
| 6 | Colin F. Camerer | 49 |
| 7 | Alan Grafen | 61 |
| 8 | Peter Hammerstein | 75 |
| 9 | Sergiu Hart | 97 |
| 10 | Ehud Kalai | 109 |
| 11 | David M. Kreps | 121 |
| 12 | Herve Moulin | 137 |
| 13 | Rohit Parikh | 145 |
| 14 | Ariel Rubinstein | 157 |
| 15 | Larry Samuelson | 165 |
| 16 | Thomas C. Schelling | 185 |
| 17 | Brian Skyrms | 189 |

| 18 Robert Sugden | 191 |
|---|---|
| 19 H. Peyton Young | 203 |
| About the Editors | 217 |
| About Game Theory: 5 Questions | 219 |
| Index | 220 |

# Preface

*Game Theory: 5 Questions* is a collection of interviews with some of the most influential game theorists of the last decades. We hear their views on game theory, its aim, scope, use, broader intellectual environment, the future direction of game theory and how the work of the interviewees fits in these respects.

The emphasis is on foundational issues as well as the way in which game theory informs other fields together with its interdisciplinary applications in economics, philosophy, computer science, interactive epistemology, biology, cognitive science, behavioral science, political theory, social science, social software, linguistics etc.

Developments in game theory have often sparked formulations of, and solutions to, central problems in the above mentioned disciplines. Game theoretical insights and methods have sharpened, radicalized and extended concepts and ideas, which without the use of results and methods from game theory would not have seen the light of day. In short, game theory serves as an important fulcrum in science across the board.

───────────── ♦ ─────────────

Game theory is thriving without any advertising and thus the idea was not to produce a bill-board for game theory along the highways of science and buildings of academia. The idea was rather to produce a forum in which the giants of the field could speak their minds freely without the standard constraints of scientific rigour imposed. Posing five open and relatively broad questions turned out ot be the perfect format for this. Besides fascinating intellectual biographies, career recaps and racy memoirs, the contributions – either in the form of direct answers to the questions or as complete essays based on them – offer interesting new conjectures, pointers to work overlooked, and methodological reflections on game theory.

Given the format of this collection the purpose is not to articulate or push any particular agenda for game theory. We have also refrained from commenting on the responses to the questions.

The responses are self-contained and readable and no overarching view of the nature of game theory is lurking in the wings. The ambition is much more modest; to initialize the discussion as to how game theorists understand their own enterprise and why these luminaries decide to make game theory central to their work.

Admittedly, the decision to pursue this kind of work may in the end much be a matter of taste and personal inclination. But in addition to intellectual biographies the book does have some interesting, illuminating and most importantly *instructive* stories to tell about the methodology of game theory and its wide range of interdisciplinary consequences.

Vincent F. Hendricks & Pelle Guldborg Hansen<br>
Copenhagen<br>
April 2007

# Acknowledgements

We are indebted to Elizabeth Pando and Christopher M. Whalin for proof-reading the manuscript and to Claus Festersen for assistance when encountering LaTeX-related problems. Our gratitude also goes out to our publisher Automatic Press ♦ $\frac{V}{I}$P, in particular senior publishing editor V.J. Menshy, for taking on such an 'unusual academic project' once again. First and foremost we are greatly indebted to our contributors for bringing this volume to life in the most erudite and entertaining way.

Vincent F. Hendricks & Pelle Guldborg Hansen
Copenhagen
April 2007

# 1
# Robert Aumann

**Professor of Mathematics**

Center for the Study of Rationality

and Institute of Mathematics

The Hebrew University of Jerusalem, Israel

---

*Interviewed at the 17th International Conference on Game Theory at Stony Brook University, July 13, 2006, by Pelle Guldborg Hansen.*

Q: Professor Aumann, how and why were you initially drawn to game theory?

A: Well, the story is this: I graduated from MIT in 1955 with a Ph.D. in pure mathematics. In fact, I had already left MIT in 1954 to work for the Forrestal Research Center, which was connected to the mathematics department at Princeton, where I did my post doc. At Forrestal they were doing operations research; very, very practical operations research, not theoretical stuff at all. There I came to work on a problem about defending a city from air attack. This was being done for Bell Telephone Laboratories, which was developing a missile under contract with the Defense Department. The problem had to do with a squadron of aircraft, a small percentage of which were carrying nuclear weapons; that is, most of them were decoys. How would one respond to this? What would be the optimal strategy to deal with that kind of situation? They sent this problem over to us, the Analytical Research Group at Princeton, and the problem was assigned to me. At that time I didn't know very much about game theory at all. I had met John Nash at MIT, when he was a young instructor and I was a senior graduate student. There we became quite friendly, and naturally he told his friends about game theory: duelling and that kind of thing. I wasn't particularly attracted to it, but I listened, and some of the problems were quite interesting. When I was assigned the aircraft problem at Princeton, I remembered these conversations with Nash and then realized that it was a game theory problem. I read a little bit about game theory, thought a little bit about it, and did what I could with that problem. In fact, I wrote a report. It may be in my files someplace, but I don't know, it is more than fifty years ago. That started me thinking about game theory, and from there I went on to more theoretical things. At that time Princeton was the center of the world in game theory, so naturally I started attending some of the game theory conferences there and mixing with the game theory people and became more and more interested. Also, I read the book of Luce and Raiffa, which got me interested in repeated games. One thing led to another. By the time I went to the Hebrew University, I was already identified as a person whose major interest is game theory. So I started giving courses in game theory, and when you teach something you get interested in it yourself. That is the story: it wasn't something that I sought out, but something on which I sort of stumbled on the path.

Q: Were you working on the Folk Theorem before you went to the Hebrew University in 1956?

A: I came to Israel in October of 1956, but I was not aware of the Folk Theorem then. In fact, I was not even working on repeated games before coming to Israel. In the summer of 1957 I was at the National Bureau of Standards, which is now in Bethesda, but which at that time was still in Washington, inside the district line. Over there I was working hard on the paper on repeated games, and I was aware of the Folk Theorem by that time. I think the major part of my work on this, the 1959 paper on acceptable points, was generated at the Hebrew University.

Q: You did much groundbreaking work on formalizing the Folk Theorem. What other examples from your work illustrate the use of game theory for foundational studies and applications?

A: That's a big question; all of game theory is about either applications or foundational studies; that's all there is, right? Everything, the whole theory, is about that. But I could single out the work on interactive epistemology, and, certainly, the work on the equivalence theorem. You might look on them as being either applicational or foundational. The equivalence theorem on the competitive equilibrium of the market uses very many different concepts from game theory: the core, the Shapley value, even the bargaining set. It is foundational if you're looking at it from the point of view of an economist. There's also the idea of correlated equilibrium. That's certainly foundational.

Q: I was just talking to Sergiu Hart a few minutes ago, and he, like many others, mentioned your work on the Folk Theorem and the agreement theorem, but in particular, he emphasized those parts of your work that most beautifully couple conceptual ideas with mathematical proofs.

A: Sergiu has made a wide range of very important contributions. Most recently, he's been working on heuristic methods for reaching equilibria; in other words, not simply algorithms, but dynamic processes leading to equilibria: Nash equilibrium, correlated equilibrium, etc. However, he's also done very important work on characterizing the value, in particular, distinguishing between the Harsanyi value and the Shapley value in non-transferable utility games. He introduced the idea of potential to characterize Shapley values and the core. In fact, some of his early work was about applying the von Neumann-Morgenstern stable set idea to the formation of oligopolistic markets. The von Neumann-Morgenstern solution is fundamentally different from other solution concepts in that it does not, in the case of markets, predict competitive equilibrium. It predicts the formation of large cartels; that is, large groups of people who cooperate with each other in order to compete with other groups. Basically, it says that a large number of small traders does not guarantee free competition. If you read the *Theory of Games and Economic Behavior* of von Neumann and Morgenstern – and I think these parts of the book were written by Morgenstern – you'll find that they did not really believe in the idea of free competition, of perfect competition. They felt that the "right" solution concept, so to speak, or at least the only one that they had for cooperative games, was the von Neumann–Morgenstern solution – the stable set idea – and this really does not lead to price competition.

Q: In the introduction to the first volume of the *Handbook of Game Theory* you and Sergiu Hart discuss game theory as an umbrella, or as a unified ...

A: ... a unified field theory. Yes, I think game theory really is unique in this respect. Other disciplines, like economics, take various real-life situations, like oligopoly, monopoly, large markets, international trade, and each one of those situations becomes a problem unto itself. Economists devise methods for dealing with this or that problem, but there is no overall methodology. Game theory, on the other hand, has the advantage of defining things very, very broadly, so that the concepts that apply to game theory apply to any game. Well, in principle they apply to any game. The core, for example, might be empty in certain games, but still it applies. In fact, any interactive situation – whether monopoly, duopoly, large markets, international trade, or, to look beyond economics, elections, bio-systems of various kinds, political systems, international relations – any one of these interactive situations is a game to which you can apply the methods of game theory, the same methods. Take the nucleolus. If you apply it to a market situation, it will yield free competition, it will yield price competition, it will give you prices. But you can also apply it to elections and it will give you minimal winning coalitions and things like that. So the same ideas apply to very different contexts. In that way, it's like one theory that gives you magnetism, gives you electricity, and gives you gravitation—it's an umbrella, that's the idea.

Q: In 2005 you were awarded the Nobel Prize in Economics together with Thomas C. Schelling. In 2002 Daniel Kahneman and Vernon Smith were likewise awarded the Nobel Prize. Some people in the press took the 2002 prize as support for a transition from classical economics to experimental economics. How do you view this issue?

A: It is interesting that the 2002 Nobel Prize in Economics was awarded to Kahneman and Smith. The behavioral economist Smith became famous for verifying the claims of economic theory with economic experiments. The Nobel Committee was not making a definite statement, as some have thought, that the assumptions and conclusions of classical economic theory are incorrect. Rather, they were saying, let's at least recognize experimental methods in this. Now, one kind of interesting conclusion was reached by the psychologist Kahneman, and another kind of conclusion – the opposite kind of conclusion, really – was reached by Smith. It's true that in the ensuing years behavioral economics gained some drive, but I'm not sure that this is going to continue. A lot of voices are challenging behavioral economics. One problem is that the conclusions of behavioral economics are based to a large extent on questionnaires and polls. These are notoriously unreliable sources. What people *say* they'll do is not what they *do*. Another problem is that the conclusions are based partly on experiments with money. Though there is some incentive given, this is usually small and people don't really pay much attention to it. True, some of the conclusions are also based on market behavior. But here again, we are not talking about large effects, things that are important to people. We are talking about whether they buy something at the checkout counter that they didn't really want to buy—and who cares about that?

I don't think that behavioral economics is going to last, though I think that it's an interesting idea. For instance, I agree with the behavioral economists that people don't think about the decisions they make. Maybe they don't think about them at all, or maybe they think about them very little. But I have this idea of rule rationality as opposed to act rationality. People act on the basis of what they have gotten used to, and what has worked in the past in the kind of situation that they are in. But this, on the whole, is usually rational. That is, people evolve rules for behavior that work on the whole and then sometimes, just sometimes, they apply them when it is inappropriate to apply them. So there is some validity, some truth, in what behavioral economists say. It is important that we face the challenges that they pose, but on the whole I'm not convinced, and I'm not the only one. Ariel Rubinstein, for example, ask *him* about behavioral economics. And many other prominent people are, let's not use such a strong word as "rejecting," but many other prominent people are sceptical, very sceptical, that behavioral economics will survive the test of time.

Q: And just to make sure, empirical economics is something quite different?

A: Yes, empirical economics is very important. Empirical economics existed before behavioral economics, and will continue to exist afterwards. It's natural that we look for something that has a connection with the real world; and indeed, empirical economics often fits very well with received economic theory.

Q: If some topics have received too much attention at times, then what do you consider the most neglected topics in late twentieth-century game theory?

A: I don't have any strong feelings about a topic that's neglected. If I feel it's neglected I will work on it. Still, the coalitional theory is more important than many people think it is. There is some kind of myth that this is useless stuff; but, most of the insights of game theory in the past have been products of the coalitional theory. I'll give you just two examples. There is the work on the

Shapley value as applied to elections, various governing bodies, or systems of government like the UN, the US Congress, and so on. This has yielded a lot of insight. Now, of course, the Shapley value also yields a price equilibrium when you apply it to economic systems. So for these applications you have all the equivalence theorems, all the results that relate game-theoretic concepts to price equilibria, and that's all cooperative game theory. The same goes for the Gale-Shapley work on matching markets, which has been applied by Roth and Sotomayor and others to many real-life contexts. That's also cooperative game theory. In fact, I think in applications you have more cooperative theory than non-cooperative theory. On the other hand, there are auctions. That's an important area and auctions are non-cooperative theory. So I don't want to be misunderstood as saying that cooperative theory is the only or the most important branch. There are people that take sides. They say "I'm a non-cooperative game theorist," or "I'm a cooperative game theorist"; and "if I'm non-cooperative, I'm going to say that the cooperative stuff is useless ...," or vice versa. I don't have such a feeling of patriotism with regard to a particular branch of the theory. I work on both. Both are important. I think that maybe – if you were to say that it was being neglected – the cooperative theory deserves a little more attention. Still, we saw several presentations on cooperative theory here at the conference. It's not so bad off that I would say "neglected." Cooperative theory is alive and kicking.

Q: So no real neglected topics?

A: No, I have no complaints about the way the theory should go or has gone so far. We are looking forward to a lot more interaction between game theory and computer science and various computer systems, but I wouldn't call that a neglected topic. It is something that has to be developed, that will be developed, and is already being developed.

Q: What do you think are the most important open problems in game theory?

A: I'll mention just one that was touched on in the conference here, and that is the problem of computational costs. This is not necessarily just a game-theoretic problem. It is a decision-theoretic problem. Let me explain. When you have a problem that involves some computation, how much should you invest in the computation? If you had some idea of how long it will take to do the computation, then you'd be okay—but you don't! Trying to figure that out may involve a computation that's more difficult than the one you're talking about. On the other hand, although it's not easy to say when you're playing a game of chess how much time you should spend on a specific move, we all somehow succeed in solving the problem at hand. So how much thinking should you invest in a given problem?

This is not an easy problem. To make progress, we have to treat it as a conceptual problem. Though you may figure out how long it will take to do the computation, just figuring that out may not be worthwhile. You have to try to approach it from some new angle. It is not clear what to do in that kind of situation. It's not your classical optimization problem. In classical optimization you just say, well, we're going to solve this problem, and then you optimize the amount of gas to put in the tank or whatever, without caring how long it takes to compute. But here you are talking not about how much gas to put in the tank, but rather how much time to compute. You *don't* want to spend more time computing how much to compute. So there is a conceptual problem here and I'm not even sure what *kind* of solution we will get. It might be some kind of evolutionary thing, without precise answers. It is a difficult conceptual problem that has been around for many years, and I don't know what to do with it.

# 2
# Johan van Benthem

University Professor and Professor of Logic

University of Amsterdam, The Netherlands

Stanford University, USA

---

Why were you initially drawn to game theory?

I should say at the start that I am not a game theorist. I am a logician enamoured of game theory – of course, in a purely Platonic sense – and accordingly, I tend to idealize the object of my affections. It would be tedious for me to pronounce on questions like 'where game theory should go': although, wherever, I do like the way it walks ...

In my early student days in the late 1960s, revolution was in the air, and we did not take it for granted that what our professors (addressed by their first names from the start, and shivering every time we quoted from half-understood revolutionary German texts) told us to read was the real stuff. That is how physics students like me found ourselves taking courses in abstract mathematics – I first heard about Category Theory through the student grapevine –, or even on the stairs of the Humanities building on our way to classes in generative grammar, or the history of Vietnam. I remember coming home one day and telling my old landlady that I had just learnt that one could prove mathematically that the Dutch language had infinitely many sentences. She looked at me strangely, and said "Johan, don't be silly". That was that. In a science faculty starved of female students, those visits of course also expressed our cravings for fashion, beauty and elegance. Eventually, those excursions took me to a class on logic, and I have been hooked ever since, switching to mathematics and philosophy. But none of these courses we took actually referred to game theory. *That* was rather a subject for books which I read in addition to our standard fare, such as Kemeny, Snell & Thompson's *Finite Mathematics* which had a tiny bit of matrix games and solutions in mixed strategies. Now the nice thing with the topology of the academic literature is that it is such a highly interconnected 'Cultural Park'. You can enter anywhere, and then there are all these wonderful trails that soon take you to unexpected new landscapes. I quickly found Luce & Raiffa's book *Games and Decisions*, which conveyed the excitement of new things going on, and the surprises of mathematical structure in what look like garden-variety human interactions. There was even the secret thrill of forbidden fruits, since someone had told me that some radicals in Sweden had proposed replacing logic by game theory as the formal apparatus which every philosopher should know. Why not at least invest a bit in this possible future world, whether or not it materialized?

A second appeal came from the cheap and immensely informative German paperback series called 'Hochschultaschenbücher' (University Pocketbooks) published in Mannheim, which contained didactic master pieces by major German professors, often even uncompromising original scientific contributions, at virtually no cost at all. These publishers were benefactors of humanity, and I am still grateful to what they did for us. One of the books I bought was Paul Lorenzen's *Logische Propädeutik*, his wonderful little monograph seeking the foundations of logic neither in the intimidating austerity of mathematical proof, nor in the lush realities of semantic truth, but rather in the structure of successful dialogical interaction. According to Lorenzen, valid arguments are those patterns from premises to conclusions in which the proponent of the conclusion has a winning strategy against any opponent granting the premises. Thus, there is a third independent pragmatic intuition of logical validity, based on viewing argumentation as a game. I have been converted to that view ever since, even though most of my professional life has been under camouflage as a model theorist, or occasionally a proof theorist. Additional evidence for Lorenzen's view came from the way in which logical operations find their natural place as operators of dialogue control in the game-theoretic setting: conjunctions and disjunctions are choices by the two players, negations are role switches. Again, this dynamic intuition just rings true to me. Nowadays, in my later years, I would also see this historically as a way of rethinking the past. Many people think that the origins of logic in Greek Antiquity come from mathematics, with Euclid's *Elements* as the paradigm of deductive proof, or the empirical sciences, with Aristotle's syllogisms as the engine of classification. But what may be more likely is that these origins lie in the debating practices of the Greek polis, for which the

Sophists trained their students, with their high-brow expression in the dialogical format of Plato's *Dialogues*. And similar issues exist elsewhere: I just heard a talk on Indian logic suggesting that its origins might lie in legal practice, i.e., again, dialogue and debate.

Even so, my love for game theory did not extend beyond teaching Lorenzen games, and later also Hintikka's evaluation games and Ehrenfeucht model comparison games, to my students – both philosophical and mathematical – as a supplement to the usual way of pouring the basics of logic into young adolescents. I did keep an eye open toward uses of games in logic and surrounding areas like the philosophy of science (Robin Giles' operational semantics for physics comes to mind), but left it to more ideological proponents of game methods in logic and argumentation theory to hold torches and make speeches. My only published effort is a little survey paper on 'Games in Logic' from 1987, in which I listed all uses of game theory in logic which I knew for the German Fraunhofer Foundation – for the sum of 1500 Deutschmarks, then a considerable amount of money for a Dutch professor. Of course, games entered my world now and then. For instance, when editing the *Handbook of Logic and Language* with Alice ter Meulen, the eventual version in 1997 lists the Hintikka-Sandu game-theoretical semantics of meaning in terms of imperfect information games in the Top Five of major paradigms in understanding natural language. But game-theoretical semantics is just a weak reflection of the richness of actual game theory, and it does not formulate a broader program. (I did write a paper in the early 1990s for a Hintikka celebration playing with a 'Church Thesis for Games', saying that, just as every computation can be mimicked by a Turing machine, every rational interaction is playable as a game.) More significant contacts between the broader epistemic logic community and game theorists had already been pioneered by then by Joe Halpern in the TARK community, by Robert Stalnaker in his work over the 1990s, and by Wiebe van der Hoek and Giacomo Bonanno in the starting of the LOFT conference series. But still, I just kept observing.

Things heated up considerably during the time of my Spinoza Award Project *Logic in Action*, when Paul Dekker and Yde Venema organized a workshop on games in 1998, where, for the first time in the history of our Institute of Logic, Language and Computation, we had a game theorist as an invited speaker, namely Arnis Vilks from Leipzig. We saw at once how congenial all this was, and how logicians and game theorists are really brothers-in-arms, or at the very least, cousins-in-arms. This also demonstrated that mutual flow of ideas was possible: from game theory into logic, as before, but also *from logic into game theory*! I started teaching a graduate seminar on 'Logic and Games' at Amsterdam and Stanford, which has been running essentially until today, resulting in various dissertations on the border line of logic, game theory, and computer science. Several of them, by Marc Pauly, Boudewijn de Bruin, and Merlijn Sevenster have already attracted quite some attention. And other talents have emerged at this fault-line between disciplines elsewhere, such as Hans van Ditmarsch, Paul Harrenstein, Sieuwert van Otterloo, or Francien Dechesne. Ever since those days, we have had a continuing series of encounters in The Netherlands, where games became a popular theme in many places. For instance, this year, we will have the 15th instalment of our informal but high-powered workshops on Logic, Games and Computation. And also at ILLC, we suddenly find that games are a unifying interest among our leading linguists, mathematicians, and computer scientists, not for profit, but for insight, and for fun! As a reflection of all this, the ILLC obtained its European Marie Curie Centre 'Gloriclass' bringing together some 15 Ph.D.-students at the interfaces between all these disciplines, while also creeping up on cognitive science now and then. By now, we also hope to take this style of thinking to a European scale, in the project 'LogiCCC' of the European Science Foundation on logics for intelligent interaction, where we are joining with like-minded people on related interfaces all across Academia. A personal research interest in games and multi-agent interaction naturally leads, at least to me, to a desire for social community building!

## What example(s) from your work (or the work of others) illustrates the use of game theory for foundational studies and/or applications?

As I said, I see two directions to the contact between logic and game theory. Let's first take the one *from game theory to logic*. I have already indicated how even just basic ideas from game theory seem congenial to notions at the very heart of logic. Many people think that interaction is just some 'nuisance' for true logic, arising from the – perhaps unfortunate – fact that we populate this planet simultaneously with many others, resulting in tons of gossip, quarrels, and cowardly compromises. (Recall that the famous Dutch logician and mathematician Brouwer was a solipsist: he heroically decided to ignore this social feature altogether.) I think, by contrast, that interaction, and the resulting 'Many Mind Problems', are just as central to logic as 'Many Body Problems' are to any significant physics. And game theory has provided notions which make this feeling precise, tying in the absolutely basic notions of truth, proof, or invariance between models with strategies in multi-player games representing different roles in inquiry. These roles range from Verifiers versus Falsifiers, Proponents versus Opponents, Duplicators versus Spoilers, or Builders versus Destroyers (this is beginning to sound like Hindu theology, but so be it). In the hands of distinguished logicians like Lorenzen, Ehrenfeucht, Hintikka, Blass, Hodges, Girard, Abramsky, Väänänen, and many others, these ideas have become powerful tools for formulating logical notions, and proving their properties. This may not be common knowledge in logic textbooks or among philosophers of logic yet, but it will. But this achievement can be appreciated in two ways. One is just as a tool, perhaps even just a metaphor. This 'Weak Thesis' is all-right: games are both tools and metaphors. But the other, more radical view is the 'Strong Thesis' that games represent something essential about logical notions, and that the two fields live in pre-established harmony. That view happens to be mine.

What new insights do we get from reformulating things this way? I will mention one, and it illustrates a non-trivial issue at the same time. Consider the major paradigm of a successful logical system, first-order predicate logic. It is replete with game imagery, once you see it properly, with Abelards and Eloises behind every tree and shrub – but let's focus on one aspect. When you view first-order formulas as denoting evaluation games, the absolutely basic issue of logical equivalence – which determines what we mean by formulas expressing 'the same proposition' – translates into the issue *when two games are the same*. Now there is no canonical answer to this, just as mathematics has no canonical answer to when two geometrical spaces are the same, or computer science to the question when two processes are the same. It all depends on natural notions of structure-preserving transformations and invariance. But at least, game theory suggests that we can look at various identification levels: global strategic forms, powers of players, or extensive forms. Correspondingly, we now get a finer view of equivalence levels for logical propositions, and we enter one of the most basic and vexing areas in the philosophy of logic. I have shown in a 2003 paper that on some views of game equivalence, predicate logic with the corresponding notion of propositional equivalence becomes *decidable*, pace Gödel, Turing, and Church. So, much can be at stake in getting clear on these matters!

Other benefits of this finer grain in logical structure are richer views of what logical constants are about. I would say that, viewed as interactive processes, games split standard logical notions into a great variety of natural notions of control. Take logical conjunction. One reading makes it a *choice* of sub-games for your opposing player, another the sequential composition of first playing one game and then the other, and a third natural reading makes it some sort of *parallel composition* of playing two activities either simultaneously, or interleaved. All this is highly congenial to the move in computer science from single Turing machines to distributed networks of computing agents, who involve finite or infinite interactions, and why by now are endowed with capacities for observation, message passing, and even goals and desires. I will not elaborate much on this computational process connection, but it is certainly another major strand in the total fabric of contacts that I am describing. We may not have the totally crystallized definite view of the natural repertoire of game equivalences, and matching logical constants here (with apologies to those who think they *have* given the world just that ...), but, praise be to the game perspective, what a much richer conceptual world for a logician to live in!

Now here is the standard objection to all this. It may be games, but is it *game theory*? After all, real game theory is about agents who have preferences and goals, who attach values to outcomes— and its major mathematical results are about appropriate notions of strategic equilibrium, and when we can have them. Indeed, that mathematical theory revolves around mixed strategies and probabilistic considerations which may look alien to pure logic. I agree that most of this structure has not found its way into logic yet, with a few exceptions here and there, in evaluation games with imperfect information, and some process theories that allow for preferences between transitions of a system. But I see no reason at all why these perspectives could not be brought in. For instance, goals and preferences become essential once you try to understand the drift of real argumentation, winning debates, or just dispensing procedural justice as a chairman. I see these phenomena as major challenges to logic, since we want to interface our accounts of validity with rational ways for making our views prevail or: for standing refuted when we should be. I have written some papers on 'winning debates' where a mixture of standard logic and real games is of the essence. Likewise, I have argued in print that mixed strategies in evaluation games make perfect sense as probabilistic mixtures of Skolem functions once you make the move from a deterministic to a probabilistic universe of objects, as happens in quantum mechanics.

Indeed, once you take this view, it will crop up in other places, too. Take, not argumentation, but the much-studied phenomenon of *belief revision*. Even though this is usually cast as a single agent recording incoming information and adjusting beliefs, in reality, it is first and foremost a multi-agent phenomenon, where we have to merge information from different sources, and where interactions with other minds make us change ours. Clearly, the eventual theory of belief revision must be about revising our goals as well, and again, values, preferences, and longer-term strategic interaction will be of the essence. This point was also made in the context of learning theory by Kevin Kelly, who shows that only in this way, can one compare different revision strategies as to their success toward stated goals. Actually, it seems to me that the *linguists* are ahead of the logicians here right now. In the work of Lewis, Parikh, Jaeger, van Rooij, Gärdenfors, and others, the crucial function of language is communication, and stable meanings emerge as equilibria in formal, but somewhat realistic, coordination games.

Now, let's look the other way, and go from *logic to game theory*. Here some of the earlier themes return, but now with a reverse thrust. And some new ones get added, since we are now looking at general games with the tools of logic. But of course, the real situation is that of a meeting of disciplines with ideas flowing both ways. For instance, the study of games is a natural continuation of the study of *process structure*, which I see as one of the major 'cultural' contributions which computer science has made to the academic landscape. Thus, game theory is giving us ideas about interactive multi-agent processes, taking it to 'the next level'. But of course, one has to merge the respective insights and modus operandi. Game theory is mainly about global notions like strategic equilibrium in a game, and it has been amazingly successful in getting away with this high-level abstraction, and extracting useful and insightful information from it. Indeed, in early stages, it even seemed as if these notions, and techniques for finding them like Backward Induction or more sophisticated fixed-point theorems, were writ in stone. Right now, I would say that this global representation needs *fine-structure*, of the sort that can be provided by ideas from both computational and philosophical logic.

The 'dowry' from *computational logic* is its sophisticated thinking in terms of process equivalences, such as bisimulation, and their matching logical languages: modal, first-order, or still more expressive, describing the corresponding invariant properties of games. I have developed these themes in my 2003 JoLLI paper 'Extensive Games as Process Models' pointing out to which extent modal and dynamic logics can then provide an explicit fine-structured account of games, and very importantly, of players' *strategies*: perhaps the real, but somewhat unsung, heroes of game theory. I think that we are at the threshold of a merge between game-theoretic equilibrium mathematics and process logics and temporal logics. In this way, we will develop a greater sensitivity to the *balance* between expressive power and computational complexity of essential properties of games, a balance which permeates so much of computational logic.

This mix and this balance become even more delicate, when we add players with limited powers of observation: just like us. In that case, we also need the dowry of *philosophical logic*, and its accounts of knowledge, belief, and other relevant informational states. To some, this seems like a strange and unhappy mixture. Computational logic is hard-core, and almost as respectable as straight mathematical logic in the foundations of mathematics. By contrast, philosophical logic is about attitudes of fallible agents with notions referring to their individual idiosyncracies: in short, the world of imperfection, compromise, and often mere mathematical bubbles. Even so, this *is* the world of intelligent agents, and we had better use all available tools to understand them. What we are seeing now is the emergence of all sorts of theories which merge ideas from computational and philosophical logic. My own area of *dynamic-epistemic logic* is a typical example. The work there on actions of information change, belief revision, and even preference change, seems to fit seamlessly with the study of games. In my 2002 paper 'Games in Dynamic-Epistemic Logic' I give several examples of this, showing e.g., how uniform strategies are exactly the ones definable by means of 'knowledge programs'. I elaborate another strand in my 2004 paper 'Rational Dynamics', providing a new epistemic take on game-theoretic 'solution algorithms', taking them seriously as processes of inner deliberation and knowledge update. My eventual hope would be that, in this way, by operating on such a broader front, we can also systematize the game theory of solution concepts and their epistemic characterizations, which has evolved since Aumann's pioneering work by the great game theorists of the 1980s. At present, it consists largely of a haphazard collection of, admittedly famous, notions and results.

All this is mainly still cooking right now, but I find this whole research area liberating. For instance, at the moment, I find myself working with my students on logics of *preference*, long considered a stagnant malarial backwater of logic. But now, we are taking the richer perspective suggested by games, inspired by the problems of Backward Induction, not as needing a 'quick fix' once and for all, but as a starting point for an in-depth dynamic analysis of preferences. We ask what leads to the preferences that we have, and how they might change dynamically under pressure of suggestions, commands, or observations of merits of other players. Van Benthem, van Otterloo & Roy 2005 gives a first example – but much more is to come.

Once again, some logicians think that this is 'dirty' or at least 'messy'. Game theory imports *economics*, and hence thinking in terms of cost, value, and so on. By contrast, I think the latter is an essential and general intellectual perspective with its own intuitions and reasoning styles, which works across academia, enriching (excusez le mot) other disciplines which it touches.

## What is the proper role of game theory in relation to other disciplines?

This question is not for me to answer. I even find it sounds too much like those old German discussions of the proper place of disciplines in some grand intellectual order of things in the late 19th century. (It is usually the working classes which need to understand their 'proper roles': the rich are free.) I think disciplines should thrive and influence other disciplines, by caring as little as possible about academic hierarchies, or who is supposed to be the guardian of what. Indeed, if I were to say anything more – which I will now proceed to do – I find logic and game theory very similar in their academic roles. Both provide very general models for analyzing intelligent behaviour, though focussing on different levels so far: logic by and large more micro, game theory more macro. For both, it is hard to say to which extent they are normative or descriptive (maybe that distinction has become tedious anyway), and their relationship to experimental cognitive science is delightfully tortuous. And finally, both do not just analyze given behaviour, they also provide for design of new styles of behaviour that can be incorporated into our human repertoire. They would really make a good match when paired as academic disciplines—but maybe, I already said that?

## What do you consider the most neglected topics and/or contributions in late 20th century game theory?

Again, this is not for me to say, and also, I do not like the term 'neglect'. Admit to it, and before you know it, some American lawyer has sued you. Of course, there are areas where I would just like to 'hear more' from game theorists. This would be in particular in *explicit theories of strategies*, richer than what we usually get, richer accounts of the *step by step dynamics* of extensive games, rather than pre-encoding everything right at the start in some huge 'type space', and finally, instead of coming up with different games for different occasions, some systematic account of *how games can change*, and what then happens to their properties, without having to re-compute everything from scratch every time.

## What are the most important open problems in game theory and what are the prospects for progress?

Here, I will just list the interfaces and developments which I expect to happen. I am seeing an emergent logic-game theoretical paradigm where logic provides the fine-structure behind the usual games, which will integrate ideas from three sources: (a) dynamic epistemic logics describing single steps of information update, belief revision, and other basic events, (b) process logics from computer science describing compositional process structure and longer-term behaviour, and (c) game-theoretic structure having to do with preference-based equilibria.

In this coming together, I expect specific teaming up between, e.g., game-theoretic fixed-point theory and fixed-point logics in computer science. Likewise, I expect exciting merges in methodology. Some people think that computational logic sits badly with real games, because its compositional methodology founders on the latter having no good notion of sub-game. This seems premature to me, since the main challenges to compositional methodology have never been easy, but the results have always been rewarding. Then, there will be a growing interface between the mathematics of dynamical systems underlying evolutionary game theory, and logics for infinite processes, perhaps even co-algebra and its modal logics as being developed today.

Note that I have cast none of this anywhere as logic being 'applied' to game theory, or game theory being applied to logic. I think those terms mean very little in significant interactions between healthy disciplines. They do not meet in order to cure each other's ailments. They meet to produce *new offspring*, and the quality of that offspring is the test of their match.

So much for technical perspectives from game theory, logic, and computer science merging into one apparatus. But there is also the arena where these frameworks will play. Based on tell-tale signs in the avant-garde literature, I expect major influences in philosophy, ranging from interactive epistemology to the philosophy of action and social philosophy. In fact, this is a safe prediction, since so much is already going on! Likewise, as linguistics is making its interactive turn, optimality theory is transforming into game theory, and again the resulting paradigm will have greater power than either component. Finally, I see all this moving into experimental cognitive science. It is very interesting to see that logic and game theory have started picking up cognitive interests around the same time in the 1990s. As cognitive scientists will see more and more that intelligent interaction is the key to understanding human rationality and success, the logic game theory connection will become stronger accordingly – and we will be scanning 'games instead of brains'.

Was this what I saw vaguely as a student reading game theory books with my pocket light under the blankets? No. But it is what I am wishing for today, and as game theory tells us, wishes can come true, provided we play our cards right.

# 3
# Cristina Bicchieri

Carol and Michael Lowenstein Term Professor
Director, Philosophy, Politics and Economics
University of Pennsylvania, USA

Why were you initially drawn to game theory?

I was a student in philosophy of science at Cambridge University in the early 80's. I was interested in Bayesian confirmation theory, but I was not happy about the formal tools available to answer questions about why we adopt a hypothesis or choose a theory. Those were the years in which Kuhn's ideas about scientific revolutions and sociological attempts to explain scientific practices were dominant in the community. I thought the social dimension of science was important, but I never believed we are dupes that respond automatically to the social environment surrounding us. Social influences should be incorporated in a model of choice, I thought, but how to proceed to do it was far less obvious. I wanted to show that there is a conventional element in the choice of which method or model to apply, but also model it as a rational choice. It was a choice, though, that did not occur in a vacuum: it had to depend upon what one expected other scientists to choose. Decision theory offered only a partial answer to my quest for a model of rational decision-making. It tells us how to make a rational choice against Nature, whereas I wanted to know what choosing rationally means when the outcome depends on what other people choose, too. Game theory gave me the answers I wanted. An article I published in 1988 summarized my views about how game theory should be applied to the study of scientific practices. It contained themes and ideas I have developed later on: social norms, conventions and common knowledge among them. At that point I had become less interested in how scientists make choices than in what it means to choose rationally in an interactive context, and whether rationality and common knowledge of rationality alone could guarantee that players coordinate upon an equilibrium. These are also important philosophical issues: epistemology contends with questions about knowledge, belief, and rationality, what they mean and how to model them. Game theory challenged philosophers to think in terms of *interactive epistemology*: what does it mean for a collective to have common beliefs or knowledge, and what the consequences of this knowledge are for the social outcomes resulting from agents' interactions. I eventually moved on to explore the relation between agents' knowledge and solutions to games. Though what drew me to game theory many years ago was a specific question about scientists' decision making, what draws me now is the recognition that philosophy cannot do without the language of game theory. In many ways, the two fields are interconnected, and can greatly benefit from each other.

## What example(s) from your work (or the work of others) illustrates the use of game theory for foundational studies and/or applications?

There are many areas of philosophy that interact in a fruitful way with game theory. An important intersection between game theory and philosophy is the 'epistemic approach' to game theory. Epistemology traditionally studies concepts such as truth, justification, knowledge and belief. Game theory usually assumes agents have common knowledge (beliefs) of the structure of the game and their mutual rationality. However, it took time before game theorists recognized that it is important to explicitly formalize the hypotheses we make about the knowledge and beliefs of the players. At the beginning of the subject, the fact that decision theory had clear foundations (for example, Savage's axioms) seemed sufficient. However, decision theory treats the agents' probabilistic beliefs as exogenous, whereas in game theory the main source of uncertainty for an agent is the way other agents will behave. If we can infer that a rational agent will behave in a particular way, then another rational agent should also predict the first agent's behavior. Probabilistic beliefs necessarily become endogenous. The epistemic approach to game theory provides a formal analysis of strategic reasoning, making explicit players' knowledge (or beliefs) about the structure of the game and the strategies, knowledge and beliefs of other players. It also has the important merit of providing an epistemic foundation for solution concepts. For example, given a certain family of games G, there will be some strategy profiles S compatible with the knowledge and beliefs attributed to the players. We can thus show that a solution S(G) captures the epistemic hypotheses E that, in turn, yield that solution.

A case in point is the backwards induction paradox. The 'paradox' arises thus: In the standard model which admits of backwards induction arguments, to determine the solution we have to begin by saying how rational players would behave at all penultimate nodes. We assume common knowledge of players' rationality, and infer that they will never get to all but one of these. Hence, since we assume that rational players know our theorem, for any one $n$ of these nodes, if a player arrived at $n$ she would know that someone is not rational. But common knowledge of players' rationality implies simple knowledge of players' rationality. So at this node the player would know inconsistent propositions. Many authors such as Rosenthal, Binmore and Reny had tried to explain what is paradoxical in these problems; my contribution was to explicitly introduce into the game the knowledge that players have about each other (hence my dubbing them *knowledge-dependent games*), and show that *limited* knowledge, but not common knowledge, supports the backwards induction solution. For example, if players have only mutual knowledge of rationality of a certain limited degree, then a player deliberating at the start cannot infer from her knowledge (because the inference would requires knowledge of more than this limited degree) that the player at the penultimate node would, finding himself there, be bewildered.

More generally, I argued (1989, 1992) that the backwards induction solution of a game is a knowledge-consistent play of a knowledge-dependent game. The reverse is also true. Thus there is an isomorphism between backwards induction and knowledge-consistent plays of certain associated games. The important point I made here is that some knowledge-dependent games have no knowledge-consistent plays. That is, if we incorrectly translate a game into the associated knowledge-dependent game, some player will be unable to reason about the others well enough (or consistently enough) to infer which action she should choose. Too little knowledge impairs deduction of the proper action, but too much knowledge is equally damaging.

Another important application of the epistemic approach is the treatment of belief revision in games. In games of imperfect information, an agent reaching an information set which had zero probability under the equilibrium strategies must have some belief about what happened in the past. Since Bayes' Rule is not applicable to updating after zero probability events, there is no obvious 'rational way' for doing this. Yet beliefs at such information sets are crucial for determining equilibrium play. To play an equilibrium, a player must know what would happen at off-equilibrium information sets. This is a crucial problem in the refinements literature, where an equilibrium is rejected as implausible if it is unstable with respect to small deviations. However, all depends upon how a deviation is interpreted by the players. A deviation is, from the viewpoint of playing a given equilibrium, a contrary to fact event. Philosophers' work on counterfactuals and belief revision is clearly important here. In particular, the work of Gärdenfors and Levi on minimum loss of information criteria allows for a ranking of refinements that is much more plausible than the host of intuitive arguments provided by game theorists to justify some equilibria as more 'reasonable' than others. Since minimum loss of information is not just a quantitative, but also a qualitative criterion, I argued (1988) that interpreting deviations as mistakes, as opposed to rational signals (whenever this interpretation makes sense), deprives rationality of its explanatory and predictive power, and therefore causes a greater loss of information.

The search for better models of agents' reasoning includes finding ways to formalize the theory of the game (T) that is used by an agent to infer its moves and therefore to compute a solution to the game. Typically, T might be formalized in a classical, first-order logic. Such a logic is sufficient to represent the structure of the game and the associated payoffs, and to infer an optimal sequence of moves relative to a given utility function. If one wants to represent the players' reasoning processes and beliefs, then it becomes necessary to use modal logics. These are, however, monotonic logics, in that any proposition $p$ entailed by a set of axioms $A$ remains entailed by any set of axioms $B$ that includes $A$. Using a monotonic logic leads to problems whenever we want to model the possibility of unexpected events occurring. For example, if an agent's theory predicts that another agent will make a certain sequence of moves, the other agent choosing otherwise is an unexpected occurrence. Such an event contradicts some of T's premises. If T is expressed in a monotonic logic, this theory, when augmented with statements to the fact that unexpected moves have occurred, becomes inconsistent. A more realistic theory of the game should be a theory that is robust in the face of deviations from predicted outcomes, i.e. it must be a theory that allows agents to play even in the presence of deviations. Antonelli and I (1995), used Reiter's default logic to formalize agents' reasoning. We specified a default first-order theory (W,D) for generic, finite extensive form games with perfect recall. Such a theory comprises two main modules or parts. The first part of the theory describes a mechanical procedure $\pi^*$ that computes the set of undominated paths through a finite tree representing a game of perfect or imperfect information in extensive form. The second part contains a set of first-order axioms W, and a set of defaults D. W includes a description of the structure of the game and behavioral axioms specifying that whenever a non-terminal node (or information set) is reached, an agent will choose exactly one among the possible moves. The defaults D represent defeasible behavioral principles to the effect that agents only choose moves allowed by recursive application of $\pi^*$. We assumed W to contain Primitive Recursive Arithmetic, which is necessary to define a function $\pi^*$ representing a particular computing (pruning) procedure. The function $\pi^*$ takes a set of nodes (or information sets) as input, and returns a set of paths through these nodes as output. In our work, we introduced a *particular* $\pi^*$ function, embodying a specific procedure for recursively pruning the tree. In general, one may want to employ different procedures on different occasions or for different purposes. Only the first part of the theory (specifying $\pi^*$) would have to be changed, leaving the behavioral axioms and the defaults unchanged.

The procedure $\pi^*$ that allows agents to recursively compute their own and other agents' undominated paths and information sets throughout the entire tree embodies a particular rationality principle: a rational player is one that only plays admissible strategies, where an admissible strategy is one that is not weakly dominated. Recursive application of $\pi^*$ along the tree embodies the concept of iterated elimination of weakly dominated strategies. Our behavioral axioms did not contain an explicit definition of rationality. However, such an assumption is already implicitly made in attributing to players the capability of computing $\pi^*$ and choosing according to it along the tree. When recursive application of $\pi^*$ returns a unique path as the solution of the game, we proved that the solution corresponds to a Nash equilibrium. Moreover, the solution concept we proposed rules out all those Nash equilibria that contain weakly dominated strategies. Building on the earlier work with Antonelli, Oliver Schulte and I (1997) proposed a new, more general version of the pruning procedure, one which provides a formal definition of agents' *common reasoning about admissibility*. We obtained several interesting results:

- Our definition of common reasoning about admissibility coincides with order-free elimination of weakly dominated strategies in the strategic form;

- In the extensive form, a strategy may prescribe choices in parts of the tree that will never be reached if that strategy is played. If we evaluate strategies only with respect to information sets that are consistent with them (i.e., those that can be reached if the strategy is played), we are led to the concept of *sequential proper admissibility*. A strategy is sequentially properly admissible in a game tree just in case it is admissible at each information set consistent with the strategy. The strategies that are consistent with common reasoning about sequential proper admissibility in the extensive form are exactly those that are consistent with common reasoning about admissibility in the strategic form representation of the game. Thus the solution given by common reasoning about admissibility does not depend on how the strategic situation is represented.

- We defined a *credible* forward induction signal as a signal consistent with common reasoning about sequential admissibility. If we allow agents to consider only credible signals, common reasoning about sequential admissibility yields typical forward induction solutions in games of imperfect information.

- In games of perfect information, common reasoning about sequential admissibility yields typical backward induction solutions. Note that the recursive pruning procedure does not start at the final nodes. Our procedure allows agents to consider the game tree as a whole and start eliminating branches anywhere in the tree by applying iterated admissibility, and therefore it does not follow Zermelo's backward induction algorithm. For example, suppose that in a game tree a move $m$ at the root is strictly dominated by another move $m'$ at the root for the first player. Our procedure rules out $m$ immediately, but the backward induction algorithm eliminates moves at the root only at the last iteration.

One advantage of this approach is that of providing a unified treatment of several solution concepts that were previously held to be different, if not incompatible. Thus a unique mechanical procedure that embodies common reasoning about admissibility can be applied in a wide variety of games. Another advantage is that rationality, and common reasoning about rationality, need not be explicitly defined. They are embedded in the mechanical procedure an agent is provided with. Much work remains to be done in this area, especially important for applications to distributed AI, where keeping the procedure and the axioms separated may present an advantage.

## What is the proper role of game theory in relation to other disciplines?

Game theory, though it is extensively used in economics and other social sciences, as well as in computer science, biology and philosophy, is an autonomous discipline. It applies to all situations in which decision makers interact and the outcome depends on what the parties jointly do. Decision makers may be people, firms, political parties, animals, robots and even genes. When firms compete for market share, politicians compete for votes, jury members have to decide on a verdict, animals fight over prey or genes compete for survival, we have a strategic interaction. Game theory is the formal language in which we model what all these interactions have in common. Yet it would be wrong to think that, since it is similar to a formal language, game theory only lends itself to a precise, skeletal representation of properties that are already there, though expressed in a less formal way. The role of game theory is that of a *model*: it gives us an idealized version of the phenomena we study, but it also leads us to explore particular facets of such phenomena.

Using a model, even when it is a formal model (as opposed to, say, a physical one), brings about new inferences, suggests new properties and in a sense *changes* the thing or process that we model. The kind of tools game theory gives us are apt to change the way we understand the phenomena we model with those tools.

As an example, think of the role game-theoretic models are playing in ethics and political philosophy. These disciplines deal with moral rules, social contracts, conventions of justice and the like, all concepts that can be given a precise meaning as equilibria of repeated games. Non-cooperative game theory is an invaluable tool that has been little understood in philosophy. And so is evolutionary game theory. Brian Skyrms, for example, has done seminal work showing how moral norms can evolve, and what their subsequent dynamics might be. Ken Binmore, Robert Sugden, Peter Vanderschraaf, Jason Alexander and myself have applied game theory to the evolution of norms, as well as to more classical problems in political philosophy. David Lewis and Edna Ullman-Margalit were the first to see the potential of game theory to explain conventions and to differentiate them from other rules, as well as to link game theory with the philosophy of language. In fact, a new and exciting area of application of game theory to philosophy is the study of how meaning can emerge. In all these areas, game theory has helped to sharpen our intuitions, allowing for 'rational reconstructions' of difficult concepts and an explanation of how social contracts, norms, conventions, values and even meaning can emerge out of various forms of interactions among agents who did not plan or expect such results.

Finally, let me mention the results that experimental game theory brings to bear on the development of ethical theories. Experiments on Ultimatun, Dictator, Trust and Social Dilemma games are helping us understand how people form fairness judgments, the cognitive dynamics involved in the process, and what drives 'fair' behavior on one occasion and dampens it in another. These are important steps that any philosopher should take in the direction of building better normative theories. Naturalizing ethics does not mean reducing what ought to be done to what is in fact done: this would be a trivial naturalistic fallacy misstep. What instead needs to be done is build our normative theories upon the solid foundation of what we know individuals *can* in fact do, and this is a whole different project. I have embarked on this project long ago, by trying to show that our ethical norms are just collectively defined and supported social norms. Some such norms are more entrenched than others, but the cognitive processes underlying norm-following, and the biases we all face in filtering and processing the social information that will ultimately decide whether or not we act in a pro-social way, are essentially the same. Without knowledge of such cognitive processes, and the behaviors they engender, ethics is condemned to remain an abstract and fairly useless endeavor.

As an example of the multifaceted use of game theory in developing better philosophical theories, consider building a theory of social norms (Bicchieri 2006). In order to provide a testable, operational definition of social norms, one has to define them in terms of conditional preferences and beliefs, and show that norms, when they exist and are followed, transform a mixed-motive game such as a prisoner's dilemma or a trust game into a coordination game of which the norm is a salient equilibrium. When we encounter a new situation, we must decide whether to obey the norm or act in a selfish way. It is as if we are playing a Bayesian game in which we assess the probabilities that the opponent is selfish or norm abiding. The theory of norms I propose predicts that, if the right kinds of expectation are present, most subjects will follow whatever norm is relevant to the decision situation. We can set up behavioral experiments in which we manipulate expectations and test this hypothesis. Clearly a game-theoretic model shapes the way we address these questions, and directs us to interesting new solutions.

## What are the most important open problems in game theory and what are the prospects for progress?

There are several areas of game theory, such as cooperative games, that have languished for some time. Here however I want to concentrate on the applications I just mentioned, since a lot more need to be done in these areas.

**DAI applications.** The epistemic approach is crucial in applications to distributed artificial intelligence (DAI). DAI focuses on solutions of problems by a multi-agent community, such as distributed planning systems, or Web agents that retrieve information for their users. Since agent performance is more effective if interactions with other agents involve coordination or cooperation, any designer of agents that act in a multi-agent environment faces the problem of encoding a strategy of interaction with other agents. Traditionally game theory has not been concerned with agents' design, and only relatively recently has explicitly dealt with formal models of agents' strategic reasoning. Thus, even if game theory can be extremely useful in providing us with methods for proving properties that are useful to adopt for designing agents, there is still a lot of work left in order to adapt these methods to the design of artificial agents. There is a need to focus on the reasoning processes of the individual players rather than on the framework within which their encounters take place. When adapting game theoretic models to a DAI environment, the choice of a strategic model applicable to the specific DAI problem must include, among other things, the development of techniques for searching for appropriate strategies that will enable agents to reach an equilibrium. Artificial agents should be *programmed*; mere identification of a solution is not sufficient. To derive a solution an artificial agent must possess *reasoning* capabilities, an algorithm that will determine its strategic behavior given the information the agent is endowed with. If agents use their available information to reach a conclusion as to how to play, this information must be explicitly represented. Often even the simplest strategic interaction has several possible solutions (*equilibria*). If agents are the product of different software designers who do not share a common protocol, giving each agent the ability to reason to a solution, and heuristic rules to interpret other agents' behavior, becomes imperative. Moreover, when dealing with artificial agents, the *complexity* of deriving a solution becomes both measurable and crucial; the computationally intractable (or at least impractical) assumptions of omniscience and common knowledge (Parikh 1987, 1995) must be relaxed and replaced with more realistic implementable assumptions.

**Experiments.** Experimental game theory has seen a remarkable growth in recent years. Experiments show that the usual auxiliary hypotheses about agents' selfish motives have to be changed, at least in many cases in which pro-social behavior is involved. Unfortunately, though many new utility functions have been proposed to explain what we observe in experiments, we still lack utility functions general enough to subsume a variety of results and specific enough to allow for meaningful predictions. Moreover, there are many open questions about the role of emotions in decision-making, and their relation to beliefs and expectations. New research done in neuroeconomics might shed light on these issues, but I believe we still need behavioral studies to assess the role of expectations, measure them, and see how manipulating expectations may lead to dramatic behavioral changes.

**Evolutionary models.** In this area, too, there is a lot of work to be done. Traditional replicator dynamics models are not adequate to model cultural evolution, and we need to develop more sophisticated imitation/learning models that take into account psychological factors. I expect more work will be done in integrating the results of lab experiments into better, more realistic evolutionary models. For example, endowing agents with utility functions that represent more accurately their motivations will allow us to build evolutionary models of, say, the emergence of institutions such as social norms that have greater explanatory value.

## References

C. Bicchieri, "Methodological Rules as Conventions," *Philosophy of the Social Sciences* 18, 1988: 477–495.

C. Bicchieri, "Strategic Behavior and Counterfactuals," *Synthese* 76, 1988: 135–169.

C. Bicchieri, "Self-Refuting Theories of Strategic Interaction: A Paradox of Common Knowledge," *Erkenntnis* 30, 1989: 69–85.

C. Bicchieri, "Knowledge-Dependent Games: Backward Induction," in C. Bicchieri and M.L. Dalla Chiara (eds.) *Knowledge, Belief, and Strategic Interaction*, Cambridge University Press 1992.

C. Bicchieri and G. A. Antonelli, "Game-theoretic Axioms for Local Rationality and Bounded Knowledge," *Journal of Logic, Language and Information* 4, 1995.

C. Bicchieri and O. Schulte, "Common Reasoning about Admissibility," *Erkenntnis* 45, 1997.

C. Bicchieri, *Rationality and Coordination*, Cambridge University Press, 1993; Second edition, 1997.

C. Bicchieri, The *Grammar of Society: the Nature and Dynamics of Social Norms*, Cambridge University Press, 2006.

R. Parikh, "Logical omniscience," in *Logic and Computational Complexity*, Ed. Leivant, Springer Lecture Notes in Computer Science no. 960, (1995) 22–29.

R. Parikh, "Knowledge and the Problem of Logical Omniscience," in Zbigniew W. Ras and Maria Zemankova, eds., *Methodologies for Intelligent Systems*, pp. 432–439. Proceedings of the Second International Symposium, Charlotte, NC, October 14–17, 1987, Methodologies for Intelligent Systems. New York: North-Holland, 1987.

# 4

# Ken Binmore

Emeritus Professor of Economics

University College London, UK

---

Why were you initially drawn to game theory?

My general views on game theory are to be found in my new textbook *Playing for Real* (Binmore [2007]) and so I plan to answer the questions from a personal point of view.

I used to invent complicated games of strategy when I was kid. As a student, I wasted the taxpayers' money by playing Poker when I should have been attending lectures. But this bad behavior came to an abrupt end when I bet my entire capital on a full house that was beaten when an opponent foolishly matched my bet with three fives—but then drew the fourth five.

However, I remained sufficiently interested in games and gaming to be surprised to find no course on game theory when I moved to the Mathematics Department of the London School of Economics in 1969. It turned out that the economists were sold on the idea that game theory was a "failed research strategy". Undaunted by their lack of interest, I persuaded a new mathematical colleague who had some knowledge of what passed for game theory in those days to offer an undergraduate course. This proved popular with students and so my colleagues decided to retain the course when its teacher left LSE two years later. Responsibility for the course then devolved on me. So I took a copy of Von Neumann and Morgenstern's [1944] *Game Theory and Economic Behavior* with me on a sailing trip to France. I don't suppose I would have got very far with this difficult book if we had not been trapped in Cherbourg harbor for a week by a series of gales in the English Channel. With nothing much else to do, I read the entire book from cover to cover. When I reached Von Neumann's second model of a highly simplified version of Poker, I could not believe that so much bluffing could be rational. How can it be right for the player who opens the betting to make the maximum raise with the worst possible hand? Von Neumann's mathematics puzzled me, and so I spent an afternoon analyzing the model by methods I could understand, and it unsurprisingly turned out that Von Neumann was right! After this experience, I was totally hooked on game theory.

In consequence, I put a lot of effort into teaching game theory to LSE undergraduates at a time when game theory was at best a fringe activity for economists. It turned out that some guy called John Nash [1996] had written some great papers in the early 1950s that were essentially ignored by the economics profession. I was particularly interested in Nash's [1950] work on bargaining, to which I added some grace notes as I went along. When some of this research was circulated in a working paper, I remember receiving a letter from Roger Myerson, in which he expressed his delight in finding that he was not alone in studying bargaining theory. However, the economics profession was stubbornly resistant. In more than one seminar, it was explained to me that "bargaining is not part of economics"!

It was only with the famous bargaining paper of Ariel Rubinstein [1982] that a sea change occurred. In a bargaining model in which two impatient players alternate in making offers until an acceptance is attained, he showed that there is always a unique subgame-perfect equilibrium whose location is determined by the relative magnitude of the players' discount factors. I added the observation that the equilibrium outcome converges on an asymmetric version of the Nash bargaining solution when the time interval between successive offers is allowed to become vanishingly small.

At about the same time, other economists had discovered the relevance of the idea of a Nash equilibrium to the theory of imperfect competition. Game theory then took off like a tidal wave, and I found myself carried along for the ride.

## What example(s) from your work (or the work of others) illustrates the use of game theory for foundational studies and/or applications?

The most important application of game theory is to economics. Economists only began to grasp its potential in the late 1970s and early 1980s, but it has now totally conquered economic theory. Its most spectacular application within economics has been to the design of auctions. My own involvement came in 2000 when the British telecom auction I was responsible for designing made a total of $35 billion. After the auction, *Newsweek* magazine described me as the "ruthless, poker-playing economist" who destroyed the telecom industry, but I notice that the telecom industry doesn't seem to be destroyed at all.

Auction design is a branch of the more general subject of mechanism design. In this context, a mechanism refers to the rules of a game invented by a government or other agency that has the power to make sure that the rules are enforced. The problem for a government is that the players in the game it invents often know more about the world than the government itself. For example, the companies who might buy licenses to operate certain telecom frequencies are obviously in a better position to value the licenses than the government who is selling them. In such situations, the government needs to delegate certain decisions to the players rather than to take them itself. For example, the government will do better by running some kind of auction than simply offering the licenses for sale at a fixed price. But the problem in delegating some decisions to the players is that the players' preferences are unlikely to coincide with the objectives of the government. Mechanism design recognizes this problem by first using the idea of a Nash equilibrium to predict the outcome of all possible games that the government might choose, and then singling out the game that generates the outcome that comes closest to fulfilling the government's objectives.

I now have quite a wide experience of applying the ideas of mechanism design in practice, and I have no doubt that it works—provided that one does not try to apply the theory outside the domain within which its basic assumptions are valid. It works in the sense that theoretical designs that succeed in the laboratory can be relied on to work in the field with reasonably high probability. It is therefore frustrating to game theorists like myself that the conservatism of most government departments should restrict the areas in which we are allowed to operate so severely.

I want next to mention applications of game theory to moral and political science. The most important result in this context is the Folk Theorem of repeated game theory, which roughly says that any stable outcome a society can achieve with the help of an external enforcement agency (like a King and his army, or God) can also be achieved without any external enforcement at all in a *repeated* game, provided the players are sufficiently patient and have no secrets from one another. Game theorists take the view that a self-policing social system must be a Nash equilibrium in which each player is simultaneously making a best reply to the strategy choices of the other players. No single player then needs to be coerced, because he is already doing as well for himself as he can. We think that even authoritarian governments need to operate a Nash equilibrium in the repeated game of life played by the society they control if they are to be stable, because popes, kings, dictators, generals, judges, and the police themselves are all players in the game of life, and so cannot be treated as external enforcement agencies, but must be assigned roles that are compatible with their incentives just like the meanest citizen. In brief, the game theory answer to *quis ipsos custodes custodiet* is that we must all guard each other.

To this insight, my own work adds a game-theoretic approach to our understanding of fairness norms (Binmore [2005]). The Folk Theorem tells us that there are many efficient Nash equilibria in the repeated games of life played by human societies. This was true in particular of prehuman hunter-gatherer societies. Evolution therefore had an equilibrium selection problem to solve. The members of such a foraging society needed to coordinate on one of the many Nash equilibria in its game of life—but which one? I believe that our sense of fairness derives from evolution's solution to this equilibrium selection problem. That is to say, metaphysics has nothing to do with fairness—if evolution had happened upon another solution to the equilibrium selection problem, we would be denouncing what we now call fair as unfair.

I go on to argue that our sense of fairness is like language in having a genetically determined deep structure that is common to the whole human race. I then give reasons why one should expect this deep structure to be captured by Rawls' original position. The question then arises as to whether Rawls [1972] or Harsanyi [1977] are correct in their opposing analyses of rational bargaining in the original position. With the external enforcement assumed by both, the answer is that Harsanyi's *utilitarian* conclusion is correct. Without external enforcement of any kind (so that there are no Rawlsian "strains of commitment" at all), I come up with something very close to Rawls' *egalitarian* conclusion. That is to say, although Harsanyi's analysis was better than Rawls', Rawls had the better intuition.

My analysis of our sense of fairness will doubtless be thought naïve by future scholars, but it is hard to conceive of a future approach that will not have a similar game-theoretic foundation.

Finally, I want to observe that attempts to provide firm foundations for game theory have profound implications for a whole range of related disciplines. Such attempts fall broadly into two classes, which I call eductive and evolutive.

Eductive game theory embraces all attempts to model players as ideally rational agents. This approach has generated numerous spin-offs, of which the most important is the theory of common knowledge proposed by Aumann [1976], who was also a major contributor to the theory of repeated games.[1] My own attempts to make progress in this area center on how to adapt theories of knowledge when the thinking processes of the players are algorithmic—so that it is no longer assumed that a rational player can decide the undecidable. It is then no longer possible to speak of *perfect* rationality as this term is currently understood. My most recent paper on this subject is pure epistemology, and the scope for making further progress in this direction seems to me enormous (Binmore [2006]).

Evolutive game theory includes all theories that model the players as less than ideally rational entities who find their way to an equilibrium by some process of trial-and-error adjustment. This process may involve individuals learning separately, or it may be a cultural phenomenon in which imitation is the most important factor, or it may be biological (in which case one usually speaks of evolutionary game theory). Evolutive game theory is too large a subject to assess here, but it will perhaps be enough to draw attention to its huge success in evolutionary biology since Maynard Smith's [1982] ground-breaking *Evolution and the Theory of Games*.

## What is the proper role of game theory in relation to other disciplines?

I think that game theory should be regarded like other mathematical disciplines as a tool without substantive content. Those critics who denounce game theorists as soulless followers of Machiavelli whose aim is to teach selfish monsters how to bring their

---

[1] Philosophers think Aumann's credit should be shared with David Lewis [1969], but Lewis's failure to produce an operationally useful definition of common knowledge is evident in the fact that the claims he makes about what needs to be common knowledge for a convention to work are mistaken.

power to bear totally miss the point. In spite of what the new school of behavioral economics say, there is no "selfishness axiom" in game theory (nor in neoclassical economics). Game theory assumes nothing whatever about what a player is trying to achieve. It tells a player how best to achieve his objectives, whatever they may be. Just as $2+2=4$ is the same for St Francis of Assissi and Attilla the Hun, so is game theory. Attila the Hun may use both arithmetic and game theory for bad purposes, and St Francis for good purposes, but arithmetic and game theory remain the same however they are used.

## What do you consider the most neglected topics and/or contributions in late 20th century game theory?

The evolutive approach to the equilibrium selection problem requires a concerted attack on the problem of how humans *learn*—both as individuals and in groups. There is a large theoretical literature, but we need an approach in which both theory and experiment advance hand-in-hand. The second area that needs serious attention is rational decision theory. Savage [1956] said that it would be "preposterous" and "ridiculous" to use the Bayesian decision theory (to which he put the finishing touches) outside a small world. However, that is exactly what economists do. But what shape should rational decision theory take in a large world?

## What are the most important open problems in game theory and what are the prospects for progress?

My answer to this question is the same as to the previous question. As for the prospects of progress, I think we are so close to breaking through in each case that it is just a matter of how many researchers choose to give serious attention to the problems.

## References

Aumann, Robert [1976]. "Agreeing to Disagree," *Annals of Statistics* 4, 1236–1239.

Binmore, Ken [2005]. *Natural Justice*, Oxford University Press, New York.

Binmore, Ken [2006]. "Can Knowledge Be Justified True Belief?," to appear in a volume honoring John Bell's sixtieth birthday.

Binmore, Ken [2007]. *Playing for Real*, Oxford University Press, New York.

Harsanyi, John [1977]. *Rational Behavior and Bargaining Equilibrium in Games and Social Situations*, Cambridge University Press, Cambridge.

Lewis, David [1969]. *Conventions: A Philosophical Study*. Harvard University Press, Cambridge, MA.

Maynard Smith, John [1982]. *Evolution and the Theory of Games*, Cambridge University Press, Cambridge.

Nash, John [1950]. "The Bargaining Problem," *Econometrica*, 18, 155–162.

Nash, John [1996]. *Essays on Game Theory*, Edward Elgar, Cheltenham, UK.

Rawls, John [1972]. *A Theory of Justice*, Oxford University Press, Oxford.

Rubinstein, Ariel [1982]. "Perfect Equilibrium in a Bargaining Model," *Econometrica*, 50, 97–109.

Von Neumann, John and Oskar Morgenstern [1944]. *The Theory of Games and Economic Behavior*, Princeton University Press, Princeton.

# 5

# Adam Brandenburger

J.P. Valles Professor
Stern School of Business
New York University, USA

---

Why were you initially drawn to game theory?

I first saw game theory in a lecture given by Frank Hahn at Cambridge University, where I was an undergraduate. Hahn drew a $2 \times 2$ coordination game on the blackboard (see Figure 1) and explained that while the $(2,2)$ outcome was the 'obvious' one in this game, the $(1,1)$ outcome was also possible. If Ann believes that Bob will play $R$, she will optimally play $D$. If Bob believes that Ann will play $D$, he will optimally play $R$. The outcome of the game depends on what the players believe about the game, not just on the 'material' payoffs.

|   |   | Bob | |
|---|---|---|---|
|   |   | $L$ | $R$ |
| Ann | $U$ | 2, 2 | 0, 0 |
|   | $D$ | 0, 0 | 1, 1 |

Figure 1

Much later, I found the wonderful passage in von Neumann and Morgenstern [18, 1944, p.42], which, I think, says the same thing (admittedly, in the context of cooperative rather than noncooperative game theory):

> [W]e shall in most cases observe a multiplicity of solutions. Considering what we have said about interpreting solutions as stable "standards of behavior" this has a simple and not unreasonable meaning, namely that given the same physical background different "established orders of society" or "accepted standards of behavior" can be built....

Having come from a science background, I was intrigued by the idea of this dependence on the 'ethereal'—on the standards or beliefs to which people adhere. (In game theory, the dependence can even be on what people believe that other people believe, and so on.) Moreover, mathematics could be used to talk about this idea.

Of course, it has taken a long time for researchers to build a systematic game theory – nowadays called the epistemic approach to game theory – that incorporates these ideas. I have written two surveys – one in 1992 ("Knowledge and Equilibrium in Games" [4, 1992]) and one recently ("The Power of Paradox: Some Recent Developments in Interactive Epistemology" [5, to appear]) – which are my attempts to describe this intellectual journey.

## What example(s) from your work (or the work of others) illustrates the use of game theory for foundational studies and/or applications?

Game theory is well suited to foundational work in a variety of fields. In my work, I have concentrated on the foundations of these foundations, so to speak. This work asks what sounds like a very classical question: What is rational behavior in a game—where each player thinks about the rationality of the other players, and so on? In fact, we still don't have a complete answer to this question.

In "Lexicographic Probabilities and Choice under Uncertainty" [3, 1991], Larry Blume, Eddie Dekel, and I developed an extension of probability theory, designed to tackle the following aspect of the rationality question. In the game of Figure 2, $R$ is inadmissible (i.e., weakly dominated) for Bob. If Ann thinks that Bob adheres to an admissibility requirement, then, presumably, she should put probability $0$ on $R$, and so will rationally play $U$. But admissibility seems to require that Ann put positive weight on both $L$ and $R$. (This is because of the standard equivalence for finite games: A strategy $s$ for Ann is admissible if and only if there is a full-support measure on Bob's strategy set under which $s$ is optimal.) If Ann puts sufficient weight on $R$, she will play $D$, not $U$. This is the conceptual problem with admissibility. (See Samuelson [17, 1992]; also the discussion below of my paper [6, 2006] with Amanda Friedenberg and H. Jerome Keisler.)

|      |   | Bob |     |
|------|---|-----|-----|
|      |   | L   | R   |
| Ann  | U | 2,2 | 0,0 |
|      | D | 1,1 | 1,1 |

Figure 2

Admissibility plays an important role in game theory. In applications, many games can be successfully analyzed this way. At the foundational level, admissibility is related to the very important idea of invariance (Kohlberg and Mertens [12, 1986, Section 2.4]). In [3, 1991], Blume, Dekel, and I proposed a solution to the admissibility problem that involves "lexicographic probability systems" (LPS's): Ann has a sequence of probability measures corresponding to a primary hypothesis about the game, an (infinitely less likely) secondary hypothesis, and so on. In Figure 2, Ann could have a primary hypothesis that puts weight 1 on $L$, and a secondary hypothesis that puts weight 1 on $R$. This way, Ann both excludes Bob's inadmissible strategy $R$ (because it gets only infinitesimal weight) and includes it (because it nevertheless gets positive weight).

In terms of probability theory, LPS's are related to the extension of Kolmogorov theory developed by Renyí [16, 1955].

Apart from dominance ideas (such as admissibility), the other central concept in non-cooperative game theory is, of course, Nash equilibrium. In "Epistemic Conditions for Nash Equilibrium" [2, 1995], Robert Aumann and I provided conditions for this concept. The case of pure equilibrium is immediate: If each player is rational and assigns probability 1 to the actual strategy choices of the other players, then the strategies constitute a Nash equilibrium. (We gave an example above, with the strategies $D$ and $R$ in Figure 1.) Mixed equilibrium is more subtle. For this, we built on Harsanyi's proposal [10, 1973] to turn randomization around, so to speak, and treat it instead as uncertainty on the part of other players about a given player's definite choice of (pure) strategy. We provided two results on mixed equilibria—with different conditions for two-player games and for games with three or more players.

Recently, Friedenberg, Keisler, and I have returned to the problem of admissibility ("Admissibility in Games" [6, 2006]). The problem of iterated admissibility – i.e., the iterated removal of inadmissible strategies – has remained largely open. We provide epistemic conditions involving "rationality and $m$th-order assumption of rationality" (where "assumption" is a concept based on LPS's) and "completeness" (a kind of richness condition). We also uncover an impossibility result: Under a nontriviality condition, rationality and common assumption of rationality ($m$th-order assumption of rationality for all $m$) is impossible under completeness. We interpret this result as indicating a limit to the idea that players reason about all possibilities in a game. In this sense, rationality, even as a theoretical concept, appears to be inherently limited.

If trying to summarize the epistemic program to date, I would point to a progressive expansion in the concept of a game: The classical matrix or tree has been augmented by structures that enable us, the analysts, to talk about the players' rationality, knowledge, beliefs, assumptions, etc. In the classical treatment, these components weren't treated formally and were poorly understood. Borrowing from an essay by Gray [9, 2001, p.866] in the *American Mathematical Monthly*:

> There are intuitions and representations, and the representations may not capture the intuitions.

The idea that Ann thinks about Bob, and about what Bob thinks about her, and so on, has always been a basic intuition about games. With the epistemic program, it has become possible to represent these intuitions within the formal theory.

## What is the proper role of game theory in relation to other disciplines?

In the predecessor volume to this one, Keisler [11, 2005, p.119] wrote in answer to the parallel question about the role of mathematics in general:

> I view mathematical research as exploring mathematical intuitions ... . Formal systems are used to clarify, sharpen, and communicate intuitive observations.

Game theory is often used in this fashion. In a pair of papers ("Value-Based Business Strategy" [7, 1996] and "Biform Games" [8, to appear]), Harborne Stuart and I have used game theory this way to explore some ideas in the area of business strategy.

Here, the mathematical intuitions are about notions of "competition," "competitive position," and the like. We provide some formal structure to define and analyze these notions.

A particular feature of this work is that we develop and apply a hybrid noncooperative-cooperative formalism. The noncooperative moves describe the players' strategic moves. The consequence of these moves is a particular market structure, formalized as a cooperative game. (See below for more on the cooperative model—including its use in analyzing competition.)

John Geanakoplos (Yale) once posed the following puzzle to me: In the area of business strategy, people often talk about a good strategy as "choosing the right game to play." But isn't such a choice just a move in a larger game? One answer is that the game being chosen is the cooperative game to be played, while the act of choosing the game to be played is a noncooperative move. This creates a formal distinction between playing and choosing a game, which also seems to fit well with usage in business strategy. In Keisler's terms, a formal system is used to clarify an intuitive idea.

Of course, there are many other examples in many different fields where game theory is used for a similar purpose.

## What do you consider the most neglected topics and/or contributions in late 20th century game theory?

I think that cooperative theory in general is a neglected area.

Von Neumann and Morgenstern [18, 1944, p.529] defined a cooperative game starting from a noncooperative game. (The characteristic function for a subset $A$ of players is the maximin payoff to $A$ in the associated zero-sum game between $A$ and not-$A$.) But later in their book [18, 1944, p.555 on], when they come to discuss market models such as bilateral monopoly and oligopoly, the characteristic function appears as a primitive. The image is of a 'free-form' market—without delineated moves for the players. Instead, the market is a fluid process where the value that each subset of players can create (i.e., the value of the characteristic function for each subset) determines the outcome.

The core is, perhaps, the most basic solution concept in this setting: Each subset must capture at least as much value as it creates. Aumann [1, 1985, p.53] explains that "the core expresses the idea of unbridled competition." It is striking how much insight can be got from using just the formalism of the characteristic function and the core inequalities. There are the famous core convergence and equivalence results, of course. But there are also subtle analyses of markets with small numbers—a fascinating example is Postlewaite and Rosenthal [14, 1974].

Still, I think it is correct to say that the contributions of cooperative theory have not been as widely known or taught as those of noncooperative theory. Happily, this might be changing, which I think would be a very positive development.

## What are the most important open problems in game theory and what are the prospects for progress?

I will mention what I see as an open area in epistemic game theory.

One of the motivations for the epistemic program is empirical, or at least quasi-empirical. Considerations of the "I think you think ..." kind seem very natural and basic in a game situation. Morgenstern [13, 1928, p.98] wrote about the battle of wits between Sherlock Holmes and Professor Moriarty (from *The Adventure of the Final Problem*) in exactly these terms.

An open area, then, as I see it, is making connections between epistemic game theory and empirical work, including experimental work. There is an experimental field – called Theory of Mind (Premack and Woodruff [15, 1978]) – which is very intriguing in this regard. This field examines the ability of humans (and non-humans, such as chimpanzees) to recognize that others may have different "mental states" from one's own. An example is recognizing that others may not know something that one knows oneself.

Of course, this leads to questions such as: Are people – some people? – able even to think about other people's mental states about yet other people's mental states? And so on. Epistemic game theory is a formal language that expresses these possibilities and works out implications for strategic interactions. The opportunity, then, is to add empirical content to this language.

## Acknowledgements

My thanks to Amanda Friedenberg, Rena Henderson, and Gus Stuart. Financial Support from the Stern School of Business is gratefully acknowledged.

## 5.1 REFERENCES

[1] Aumann, R., "What is Game Theory Trying to Accomplish?" in Arrow, K., and S. Honkapohja, eds., *Frontiers of Economics*, Basil Blackwell, 1985, 28–76.

[2] Aumann, R., and A. Brandenburger, "Epistemic Conditions for Nash Equilibrium," *Econometrica*, 63, 1995, 1161–1180.

[3] Blume, D., A. Brandenburger, and E. Dekel, "Lexicographic Probabilities and Choice under Uncertainty," *Econometrica*, 59, 1991, 61–79.

[4] Brandenburger, A., "Knowledge and Equilibrium in Games," *Journal of Economic Perspectives*, 6, 1992, 83–101.

[5] Brandenburger, A., "The Power of Paradox: Some Recent Developments in Interactive Epistemology," forthcoming in *International Journal of Game Theory*. Available at www.stern.nyu.edu/~abranden.

[6] Brandenburger, A., A. Friedenberg, And H.J. Keisler, "Admissibility in Games," 2006. Available at www.stern.nyu.edu/~abranden.

[7] Brandenburger, A., and H.W. Stuart Jr., "Value-based Business Strategy," *Journal of Economics & Management Strategy*, 5, 1996, 5–24.

[8] Brandenburger, A., and H.W. Stuart Jr., "Biform Games," forthcoming in *Management Science*. Available at www.stern.nyu.edu/~abranden.

[9] Gray, J., "Weierstrass, Luzin, and Intuition," *American Mathematical Monthly*, 108, 2001, 865–870.

[10] Harsanyi, J., "Games with Randomly Disturbed Payoffs: A New Rationale for Mixed Strategy Equilibrium Points," *International Journal of Game Theory*, 2, 1973, 1–23.

[11] Keisler, H.J., Interview in Hendricks, V., and J. Symons (eds.), *Formal Philosophy: Aim, Scope, Direction*, Automatic Press / VIP, 2005, 117–123.

[12] Kohlberg, E., and J.-F. Mertens, "On the Strategic Stability of Equilibria," *Econometrica*, 54, 1986, 1003–1037.

[13] Morgenstern, O., *Wirtschaftsprognose, Eine Untersuchung ihrer Voraussetzungen und Möglichkeiten*, Springer, 1928.

[14] Postlewaite, A., and R. Rosenthal, "Diasadvantageous Syndicates," *Journal of Economic Theory*, 9, 1974, 324–326.

[15] Premack, D., and G. Woodruff, "Does the Chimpanzee Have a Theory of Mind?" *Behavioral and Brain Sciences*, 4, 1978, 515–526.

[16] Rényi, A., "On a New Axiomatic Theory of Probability," *Acta Mathematica Academiae Scientiarum Hungaricae*, 6, 1955, 285–335.

[17] Samuelson, L., "Dominated Strategies and Common Knowledge," *Games and Economic Behavior*, 4, 1992, 284–313.

[18] Von Neumann, J., and O. Morgenstern, *Theory of Games and Economic Behavior*, Princeton University Press, 1944 (Sixtieth Anniversary Edition, 2004).

# 6
# Colin F. Camerer

Professor of Business Economics

California Institute of Technology, USA

---

Why were you initially drawn to game theory?

Three forces drew me to become interested in game theory.

The first was that we were taught very little about it as graduate students at the University of Chicago in the late 1970's. My price theory teacher made a brief comment about how von Neumann and Morgenstern's theory had promised to solve the problem of how a few oligopolistic firms behave, to fill in the large gap between monopoly and price-taking perfect competition, but that their theory had failed to do so. (Perhaps this was true, in the sense that it was only with the later efforts in the 1980's in the development of theoretical IO, by Fudenberg and Tirole and others, that game theory became rigorous and insightful about firm competition.) The only game theory course in economics or the business school that I recall being taught at that time was Lester Telser, who taught about applications of the core.

As a result of missing out on game theory, I was especially eager to learn about it during my first academic job at the Northwestern business school (teaching business strategy). At the time the Northwestern MEDS Department was full of many of the rising stars of game theory and its application areas in political economy – Dave Baron, Bengt Holmstrom, Paul Milgrom, Roger Myerson, John Roberts – and many who are still there, such as Ehud Kalai. It was remarkable to gather all this talent together in one place, particularly in a business school rather than an economics department. Even though I was an assistant professor, I sat in Paul Milgrom's IO course, and went to a lot of seminars to learn what was going on.

A second force drawing me to study game theory came from reading Luce and Raiffa's 1957 book *Games and Decisions*, in 1981 or so. Their book was a very unusual and powerful collaboration across disciplines, since Luce is a mathematical psychologist and Raiffa a decision theorist. They took the mathematical details very seriously, of course, but were also opinionated about elements of psychology and likely behavior. It struck me that trying to predict behavior in strategic situations had just the right degree of difficulty for social science (not impossible, but certainly not too easy); and if it was possible, the result could be extremely powerful in many domains.

The third force came as I began thinking about doing experiments in economics. Somewhere along the way, I implicitly realized that one of the hardest parts of experimentation – creating a careful design that is likely to produce a conclusive and interesting result – was already done for you in game theory. In most cases, theory provides examples (often particular numerical cases that are the subject of debate) and design details for you: The experiment just walks off the page and into the lab. So experimentation in this field is much easier than in other areas where you have to both make a lot of design decisions (how many subjects, how many periods of play or trading, what shapes of demand and supply, etc.) and defend them to critics. A ready-made design that is familiar to theorists is also easy to defend because, basically, people recognize the game and the design as something up for debate, and are typically curious about the results. (It is often hard to defend designs in areas where theory is so general that it does not specify particular parameters, such as growth economics, where we have done some experiments.)

And while game theory is an ideal subject of experimental study, it also seems that experiments provide an especially good kind of data about behavior in games. The reason is that predictions about behavior often depend so delicately on what people know, the order of moves, mutual or common knowledge about types and value distributions (e.g. in auctions), and so forth. The more assumptions are crucial to specifying a game, and the more the prediction depends sensitively on the assumptions, then the more valuable experimental control is. It is notable that there have been quite a few game theorists who took up experimentation at some point (Ken Binmore and his collaborators Avner Shaked and John Sutton, Vince Crawford, my colleague Tom Palfrey, Ariel Rubinstein, and Larry Samuelson). Their efforts show the complementarities between thinking deeply about theory and being so curious about what happens that one is motivated to go create data.

By the way, let me use this opportunity to make some critical points about generalizability from experimental data. It is absolutely crucial in talking about experiments in game theory (and in other domains of economics) to distinguish between "what experimental data have shown" and "what the *experimental method* might show". People are often curious about how well abstract game experiments, conducted over periods of hours for modest stakes with student subjects (though not always), produce regularities which would generalize to corporate decisions by highly-educated and experienced professionals, decisions by groups, learning over months or years, and so forth. It is quite reasonable to question generalizability along these dimensions of particular findings. For example, rejections by students of $10 out of a $100 ultimatum game "pie" might not tell you directly what a labor union would do in responding to an 11-th hour offer before a strike deadline.

What is extremely frustrating as an experimenter is having a particular experiment criticized on the grounds of generalizability, but having a one-sided conversation about how the results would change if the experiment was changed to be more lifelike or generalizable. For example, many times when I've given seminars people have asserted that results might conform more to rational or equilibrium predictions if the stakes were higher. Often I ask what the stake would have to be to produce equilibration. Nobody has ever given a concrete answer. Without a concrete answer – or better, of course, a theory that says precisely how financial incentives influence the amount of thinking or speed of equilibration – it is a critique that cannot be replied to.

Furthermore, even if results from certain lab paradigms may not generalize well to certain naturally-occurring situations, that fact is not a criticism of laboratory experimentation in general; it is just a criticism of previous paradigms and, indeed, is essentially a request for more experiments. So just as in any field, the kind of criticism that is most constructive and welcome is specific suggestions about what kinds of experiments would generate empirical facts that would best test a theory or generalize to a particular setting. Experimental economists now have a lot more tools, between field experiments, use of the internet for large-sample and long-lasting experiments, and the ability to reach broad subject populations, to run almost any type of experiment we want.

So what drew me to game theory was: Being deprived of it in graduate school, then landing in a hotbed of activity (like a farm boy who moves to the big city); being intrigued by the prospects for mixing psychology and game theory to make predictions; and seeing it as especially well-suited for experimental design.

## What example(s) from your work (or the work of others) illustrates the use of game theory for foundational studies and/or applications?

There are three topics from my work that go right to the foundations. One is the highly parametrized form of limits on iterated strategic thinking developed with Teck Ho and Kuan Chong, which we call the cognitive hierarchy (CH) approach. In these models, players doing $k$ steps of thinking either believe that others are doing $k-1$ steps (in Rosemarie Nagel's setup, or work by Vince Crawford, Miguel Costa-Gomes (*Econometrica*, 2001, AER 2007) and others), or that others are doing 0 to $k-1$ steps. These models fit data from one-shot games quite well and also provide initial conditions for learning models (which is important when there are multiple equilibria, since the initial conditions often predict which equilibrium a group or population will converge to). I say a little more about this below.

Another concept we have begun to explore, in a GEB paper with Meghana Bhatt, is the concept of equilibrium as a "state of mind" detectable by fMRI brain imaging. In that paper, we find that when subjects are in equilibrium – in the sense that they both accurately forecast what others do, and also best respond given their accurate beliefs – there is no distinction in brain activity when they are asked to choose strategies and to form beliefs. This is the simplest evidence that accurate prediction of others requires activation of overlapping brain regions used to make one's own choices, as if subjects must simulate the process of choice by others in order to make accurate guesses. When they are out of equilibrium there is more activity making choices than when making guesses, consistent with a CH view that disequilibrium results when players do not think hard enough (measured by differences in activity) about other players' choices.

A quite different contribution comes from thinking about development of common norms for naming pictures in a Schelling-type matching game (work with Roberto Weber, *Management Science* 2004, and unpublished work with Weber, Scott Rick, and with Lauren Feiler). Here the game is a 'team' game because players have a common interest in coming up with a mapping from visual images to semantic labels (because it costs them money if they are slow in communicating which visual image to coordinate on, and when they make mismatch mistakes). The game theory is not very interesting but what subjects do is interesting. One thing we learned is that they are very good at noticing subtle visual distinctions and finding natural-language labels for them (see also Michael Bacharach and Michele Bernasconi, GEB). In one experiment we even gave pictures of different kinds of marble textures, hoping it would take them a long time to find natural-language labels, but even there they were very good at finding names. This work gives us an empirical way to study the codes that develop in groups and companies, and to measure the productivity value of various codes and perhaps their persistence and robustness. These experimental paradigms provide a way to study what happens in companies and gives some empirical discipline to study vague concepts like culture of corporations, which seems to be extremely important but has not been studied very empirically deep.

## What is the proper role of game theory in relation to other disciplines?

I view game theory as a usefully general language for describing social interaction in many disciplines. To be sure: Game theory does not unify disciplines, so much as provide a common language. Scientists in different disciplines can then swap tips and inspiration about mathematical concepts. For example, evolutionary game theory seemingly came from biological considerations (e.g. animals are born with genetically-expressed strategies that do not vary much across their lives). But the same mathematical tools are potentially useful for thinking about learning by people or firms in economic settings. Whether a particular model, such as replicator dynamics, fits well in one setting or another is an empirical question. The point is that one discipline creates a tool that can be borrowed and put to use, perhaps in a way that is very insightful and perhaps not, by another discipline.

By providing common tools, game theory also gives social scientists a common focus to compare the power of different field-specific variables to predict what will happen in various games. Consider a game of trust. In the binary form first used by Camerer and Weigelt (1988, *Econometrica*) and later by Berg, Dickhaut and McCabe (1995, *Games and Economic Behavior*), one player can take a safe option or take a risky option. The risky option creates larger mutual gain, but the second player can choose whether to betray (returning so little that the first-moving player would have done better choosing safe initially) or repay (yielding a Pareto-improvement compared to the safe choice).

In predicting what will happen in an experiment on this game, economic game theorists are inclined to focus on structural variables; the payoffs, and whether the game is repeated. Cultural anthropologists might say that group affiliations and perhaps socialization will influence trust. Sociologists might emphasize properties of the group from which subjects are drawn, such as cohesiveness and solidarity. In the purely economic approach, language should play no role, but we know from many experiments on prisoners' dilemma that pre-play "cheaptalk" is one of the strongest variables increasing cooperation, rather than statistically worthless.

Each discipline can choose to specialize in the study of different variables, but if the goal is to be able to predict what happens then the disciplines can learn from each other. There is also hope that variables from different fields can be translated into a common mathematical language. For example, the sociological concept of a "norm" can be defined as an expected pattern of play (beliefs which are self-enforcing), such that violations of norms are punished which enforces them. Another example is that in the 1960's social psychologists thought a lot about "equity theory"– how equity of outcomes was perceived and acted upon. Similar intuitions now guide economists studying social preference models (i.e. models in which $A$'s utility depends on $B$'s outcome or beliefs), such as inequity-aversion, reciprocity, and social image. The difference is that the high-powered tools used by economists (e.g. psychological games) enable them to go beyond the intuitions of social psychology. The models then generate surprising predictions which inspire further experimentation and empirical work; those new data will surely generate new surprises and inspire more theorizing, in a cyclic dialogue between theory and observation.

## What do you consider the most neglected topics and/or contributions in late 20th century game theory?

My choice for the most neglected topic is modeling of how agents actually play games, with models guided by empirical facts and grounded in psychological detail. That is, there are a few such papers, but many, many more papers which never directly address the question of what the theory is thought to predict and whether theorized behavior is likely to occur empirically or not. There are also deep papers about rationality limits (e.g. finite-state machines and memory bounds), but they are not usually tightly constrained by empirical facts and do not generally rely on the kinds of constructs psychologists use to model actual thinking processes. These theories are useful in the way that applied mathematics typically is, but may not be the single best basis for constructive descriptive theories of individual agents playing games.

I have a hunch that game theorists became disinterested in data because of experiences like people who worked at RAND in the early 1950s. Nash himself and others did some experiments and were apparently surprised that people did not obey simple principles. Rather than regard this as a descriptive challenge, my sense is that the data were dismissed—game theory then took a normative applied math route rather than a mathematical psychology route of trying to create models of why people were not obeying the theory. Now we have a lot more ideas of why people may violate theory based on mutual rationality and best response (in the form of quantal response equilibria (QRE) and cognitive hierarchy (CH) approaches); but it is surprising that it took 50 years for the latter theories to emerge.

My hope is that teaching and research on game theory will be heavily influenced by some combination of QRE and CH approaches. The first maintains the rational expectations assumption (i.e. beliefs are correct) but allows small mistakes to be made more often than large ones; the second assumes heterogeneity and reasoning, so that some players do not reason strategically but others realize that some players are not strategic. In the form Teck Ho and I have studied (QJE 2004, drawing on earlier work by Rosemarie Nagel and Dale Stahl) players do different amounts of calculation but choose best responses. These approaches are not as easy to compute and work with theoretically as standard equilibrium concepts but are very useful in explaining anomalies observed in experimental data, and are very likely to be useful in explaining anomalies in field data too.

There is a lot of normal science to be done on QRE and CH models. Furthermore, to me it is much more natural to teach game theory by starting with the CH approach and QRE approach since it resonates with students' intuitions about how people think. Then move to Nash equilibrium as a limiting case of QRE when response to expected payoffs is extremely sensitive (and in some games, when the CH players are doing many steps of strategic thinking). Starting with QRE also introduces the notion of trembles immediately, so that problems like incredible threats simply do not arise, so you can bypass the issue of why Nash equilibrium needed to be repaired by imposing subgame and trembling hand perfection.

On the other hand, the idea of endogenizing link choice in network formation is a topic neglected in sociology that game theorists have come to study recently, in the 1990's (see e.g. Matt Jackson's World Congress 2006 paper for an accessible review of this "strategic approach to networks"). This is an important topic because network links are hugely important in many domains—job referrals, mate choice (many married people report that they were introduced to their spouses by friends or acquaintances), diffusion of innovations, corporate board interlocks (Jerry Davis at Michigan shows how poison pills and golden parachutes appeared to spread through interlocks of board directors), and so forth. Furthermore, figuring out which links will be made when link-forming resources are scarce is a task that is better left to game theorists like Jackson and Asher Wolinsky than to mathematical sociologists (who just don't have the tools and knowhow to solve for endogeneous equilibrium links). This could be an exciting area because the study of networks could be a shared focus of attention among computer scientists, statistical physicists, economists, sociologists, epidemiologists, and anthropologists. Networks are also of great practical importance. For example, if you wanted to know what simple changes might stem an AIDS epidemic, you might look at networks of transmission as sociologist Martina Morris has. If you wanted to know how or why companies struggle or succeed, you might "x-ray" their internal networks of influence (the real organizational chart, not the official one) as business sociologists do. Most of the work in this area simply takes a snapshot of a network in time and catalogues network properties (perhaps correlating network features with some outcome in a cross-section). Endogenizing networks and studying their evolution over time as a function of conscious choice and learning by agents is something game theorists are well-suited to study.

### What are the most important open problems in game theory and what are the prospects for progress?

The open problems that interest me are not the deep technical ones other contributors might write about (and are much better qualified to write about than I am); these problems reflect my own interests in limits on rationality and psychology.

One interesting question is mental representation of strategic interactions. People actually play games in their minds, or occasionally with an external representation (e.g., a war room in war games). Rarely do they look at a matrix or tree and reason about it the way game theorists do. This is especially true for animals, who may have a distilled representation which is encoded purely as neural firing rates from particular motor strategies shaped by within-lifetime learning or sculpted by evolution.

The importance of mental representation shows up in the lab, when we wonder whether subjects represent or perceive the game as we do. For example, I did some experiments on stag hunt or assurance games at Caltech. I used a random protocol in which a small pool of subjects were randomly matched with others round after round. One subject persistently chose the inefficient safe strategy and that subject's behavior eventually "infected" the whole group (through the random matching) and led to population convergence on the inefficient equilibrium. The subject who drove the population to inefficiency commented in written debriefing that 'I can't believe you guys are still studying the prisoners' dilemma!'. When I explained verbally that the game was *not* a PD he was bewildered. His focus on the inefficient Nash equilibrium led him to immediately classify the game as a PD (mistakenly) and 'defect'. Larry Samuelson (*JET* 2000) has a nice formal model of this in which there is an attentional cost to perceiving the game as an unfamiliar one (similar to what cognitive psychologists call a 'Stroop task', the difficulty of inhibiting an automatic perceptual response).

I wrote a working paper about mental representation in games in 1998 but got stuck, partly because I could not interest cognitive psychologists in the problem very much so there were few data to work with. A notable exception is Philip Johnson-Laird, who wrote a short paper applying "mental model theory", used to study reasoning, to games. Giovanna Devetag and Massimo Warglien also have some nice work on this (coming out in the *GEB*), showing why some games are difficult to reason through, by representing $2 \times 2$ matrix games as bi-orders and showing that some bi-orders are conceptually more difficult than others.

Another interesting cognitive question is how learning spills over across games which are not identical, but are similar in structure or in surface features. For example, Marc Knez and I did some experiments (published in *OBHDP*) in which subjects first play an assurance or stag hunt game, and reach the efficient equilibrium, then switch to a PD. The norm of efficiency they develop in one game leads to more cooperation in the subsequent PD (compared to a counterbalanced condition in which they play the PD first).

Paul (PJ) Healy has a nice paper (being revised for *AER*) on how perceptions of type correlation can explain behavior in 'gift exchange' games in which firms prepay a wage and then workers exert costly effort which benefits the firm that hired them. The emphasis on type correlation inspires the need to figure out how type correlation works psychologically, which is roughly equivalent to the problem of developing social categories and stereotypes (e.g., if one Chinese female worker behaves in a certain way, what inferences are drawn about Chinese behavior, or about female behavior). This is a good example of how a question that is raised by generalizing incomplete information games to allow type correlation requires an answer from social cognition about which type correlations are likely to be common, and why. More understanding of this might come from 'essentialist' psychology cultural anthropologist studies, about what types of groups are considered basic and statistically reliable.

A different topic which requires some help from psychology is what might be called 'boundedly rational mechanism design'. In mechanism design a designer creates rules that are designed to evoke individual behavior to accomplish the designer's goal. Usually these problems are solved by assuming people satisfy an individual rationality (IR) constraint. However, the rationality constraint should include some concept of calculation costs or comprehensibility. That is, when the IR constraint is written down as an inequality comparing costs of two different strategies (typically, comparing an optimal response given the designer's rules with an outside-option value) it is easy to be lulled into believing that people will have the same comparison in their heads (or shaped by trial-and-error learning or imitation). But if computational costs are included, the optimal design might differ. Generally, accounting for computational costs will privilege simple rules, rules that are easily computable or learned.

As for progress, I am always optimistic about progress of any sort. The reason is that young scientists who are still in college (or kindergarten!), and new mathematical and perhaps empirical techniques, as well as computing power, will be able to do some things we cannot even anticipate. So I don't think we should ever feel constrained as a field to avoid tackling something in science which seems too hard or impossible. Learning curves are so steep (and spillovers so wide) that progress can be very rapid.

# 7
# Alan Grafen

Professor

St. John's College
University of Oxford, UK

---

Why were you initially drawn to game theory?

I need to begin the interview, before answering even the first question, by explaining to the reader why I'm here at all. The three most obvious candidates to represent biological game theory are George Price (1922–1975), who invented the modern way of applying game theory in biology; John Maynard Smith (1920–2004), his co-author in the first paper on evolutionarily stable strategies, and the major figure overseeing the development of game theory in biology, including some early seminal contributions, and writing 'the book' on game theory in biology; and Bill Hamilton (1936–2000), who had explicitly used a game theory concept earlier, and who, although he always preferred a formal framework based on population genetics, brought game theory thinking to many parts of biology. These major figures would have made more natural contributors to a volume of this kind.

My first encounter with game theory, during my school days but not at school, was in popular mathematics books, perhaps of Martin Gardner, and I enjoyed it along with many other recreational mathematical topics. But it was game theory rather than hexaflexagons, Moebius strips, or liars' paradoxes that launched me on my University career when, in 1973, I applied for admission to Oxford University. The two interviewers, economist Donald Hay and sociologist Anthony Heath, explained the Prisoners' Dilemma, showing me the payoff matrix in the form in which each of the four squares is divided into two triangles, and one player's payoff is in the upper and the other's in the lower triangle. Confused by the presentation, I insisted on writing out my own payoff matrix, with the payoffs to the column player as the only entry in each square. I then observed that Defect dominated Cooperate, and so the rational action was unquestionably to play Defect – what more was there to say? I did not over-emphasise that I had met the Prisoners' Dilemma before, and my performance clearly impressed them as I was awarded an Entrance Scholarship to read Politics, Philosophy and Economics.

The appeal of game theory was probably that it provided a very powerful and suggestive approach to understanding human behaviour for a student whose view of life had been taken over by logical positivism at about the age of 13 through reading 'Language, Truth and Logic' by A.J. Ayer (1936). This virulent philosophy is corrosive of naïve minds, and had been popular in Oxford in the 1930s, though very outmoded by 1973. Oblivious to philosophical fashion, it was attractive to me to be able to cast the logical positivist net wider by reducing sophisticated human behaviour to an amoral positivist framework. Indeed, it was easy to believe that all human behaviour would in principle succumb to the combination of game theory and logical positivism.

I must confess, however, that on the few occasions when I have tried consciously to apply game theory to real decisions in my own life, it has not helped me in the least. As a true believer, I did not take these practical setbacks too seriously to heart.

At University I quickly switched to study Experimental Psychology. My economics tutor was on sabbatical, leaving politics and philosophy. Although there was some game theory in one of the politics books I read, it was hopelessly naïve (I mainly remember the shiny silver cover), and I switched because the psychologists were doing strategic animal behaviour in their first term. There was game theory in two of the eight final courses that followed, and I was privileged to have one-to-one tutorials with Richard Dawkins, at that time writing 'The Selfish Gene' (1976). The genius of that book was to present a whole and coherent picture of adaptive natural selection in which game theory played a natural and central part. I moved on to a masters degree in Economics, in which games were studied formally. Those were golden days in Oxford for economics—my supervisor was Jim Mirrlees, and other lecturers included Amartya Sen, Christopher Bliss, and Joseph Stiglitz. Even before starting the economics course I had begun doing research on game theory in biology, and at the end of it I switched again and took a doctorate in Zoology with Richard Dawkins, in which a major element was the use of the theory of Evolutionarily Stable Strategies, and game theory in biology more generally.

Since then, my main research work has involved using game theory and optimization ideas in biology in various ways. Today, in my ongoing 'formal Darwinism project' (see alan.grafen.name.for up to date citations) I am attempting a grand unification of theories of natural selection, in which the strategic elements will be explicitly represented, and formal links proved between equations of motion and optimization programs. It could, not unfairly, be described as an attempt at a neoclassical-neoDarwinian synthesis. I estimate that program will last another five years at least, so game theory will have carried me from the beginning of my academic career through quite close to the end, and quite likely beyond. The year in which I impressed my interviewers over the Prisoners' Dilemma was the same year in which Maynard Smith and Price (1973) published the original paper on ESSs, so I was in pretty much from the beginning.

I yield to the temptation to mention one of my own papers that has made no impression whatever in biology, game theory or economics. 'Fertility and reproduction in *Femina economica*' (Grafen, 1998) was my attempt to make the simplest possible model in which humans could be viewed formally both as utility-maximising consumers and as fitness-maximising organisms. There is a kind of general equilibrium game, and I prove links between economic and biological equilibrium concepts. Not surprisingly, utility has to be defined in terms of offspring. The five years I spent on this appear to have been wasted, and I have been pondering what lesson to draw. My early career encouraged me in the view that I should pursue ideas in formal terms as far as I could. But two things have changed. I learned more mathematics, so that 'as far as I can go' is now very much further. It may also be the case that fashion has somewhat turned against 'high theory', and favours more low-tech, more empirical work that lacks the taint of master narrative. Thus, my current quandary is how far to be seduced by fashion, and how far to remain faithful to the original intellectual interest that first drew me to game theory.

## What example(s) from your work (or the work of others) illustrates the use of game theory for foundational studies and/or applications?

The central point in biological game theory is how to identify the agents and maximand. An early attempt by Lewontin (1961) had considered species as agents, and the avoidance of extinction as the goal, but this led nowhere. The breakthrough by George Price, published as Maynard Smith and Price (1973), was to identify the individual as the agent, and fitness as the maximand. The constraint set was what an *individual* was able to do. The glory of this formulation is that it represents so directly Darwin's view that natural selection brings about design. The ability to capture design mathematically permitted biologists to engage in what, otherwise, would have been regarded as unwarranted anthropomorphism. Formal optimisation ideas had been around in biology since Fisher (1930), but had run into difficulties of misunderstanding. What allowed game theory to avoid the unjustified but harsh rejection by mathematical population geneticists accorded to the Fundamental Theorem of Natural Selection (Fisher 1930)? My answer is that it didn't avoid it, but rather it survived it through not dealing with genes, thus placing itself at a distance from the central technical competence of population geneticists. Only by avoiding one set of foundational questions was game theory able to survive and pose a different set of foundational questions.

What is the central point made by game theory in biology? It is that social interactions can in principle be explained by individual pursuit of a simple goal, namely having as many offspring as possible. (I simplify somewhat.) The Hawk-Dove game has little if any empirical relevance (organisms just ain't that simple, and notably are under no obligation to become simple while we develop our methodologies). The main conceptual points it was used to make (mixed versus pure strategies; symmetric versus asymmetric games; pairwise contests versus 'playing the field'; conditional strategies) were conceptually very simple indeed, but to have and to hold these ideas it seemed necessary to have formalism. And with the formalism went theorists who constructed more complicated theories, of variable value. But the basic ideas became a required part of the mental furniture of biologists who study behaviour, and rightly remain so.

Turning away from the conceptual, this general strategic awareness has inspired many very significant literatures in biology. The collection of Dugatkin and Reeve (1998) points out social foraging, cooperation, contests, communication including handicaps, kin-based cooperation, sibling rivalry and parent-offspring conflict, habitat selection, predator-prey interactions, and learning. Many of these have considerable sub-branches. It is fair to say that game theory has had a weighty influence. At the empirical-inspired end of the spectrum is the work of Susan Riechert (see her 1998 paper for a review). Here the focus is on the spider, and theories are constructed for a whole set of particular issues that arise in the fieldwork. The result is an enormously richer understanding of the evolutionary pressures and ecological setting in which the spider finds itself than would have been possible without game theory. At the concept-inspired end lies the handicap principle of Zahavi (1975, 1977, Zahavi and Zahavi 1997), and my own work (Grafen 1990a, 1990b) that sets out in game theory terms a highly abstract interpretation of it. A large empirical literature continues to flow from that work, in a wide variety of species.

At a more theoretical level, there is one heroic model that was deliberately aimed at understanding general features of animal conflict in explicit game theory terms, and stands alone in its class. The 'Sequential Assessment Game' (Enquist and Leimar 1983, 1987; Leimar 1988) took ESS theory to its limit in one direction at least, and tackled the following stylized facts about animal conflicts: (i) there are a number of behavioural elements, say A, B, C, etc. that are cooperatively engaged in, during the course of a contest (ii); the contest begins with one type, repeated, say a number of As (iii); contests can finish at any stage, but as they lengthen, a new type of element is added, which is performed in a mixture with the already introduced elements. Thus a fight may begin with As, then have a phase with a mixture of As and Bs, then a further phase with a mixture of As, Bs and Cs; and so on. The assumption of the theoretical game is that each instance of an element provides noisy information about dimensions of fighting ability, with each type of element providing a different weighting of the dimensions. There is also a cost to each type of element. The explanation offered by the model is that organisms choose to engage in the element that provides the most cost-effective information, which is why a fight begins with one type of element. But repetition reduces the informativeness of each further repeat, until a second element becomes equally valuable; the two types are mixed, maintaining their equal but jointly declining information value, until a third type becomes equally informative, and so on. Eventually, the relative fighting ability becomes clear enough to both parties, and one animal withdraws. The explanatory power of this model has also been investigated empirically with great success (see Brick 1999 and references therein). It is notable, and entirely in keeping with the rough-and-ready biological tradition, that all signaling had to be ruled out somewhat arbitrarily, so as to render the model tractable. While there are aspects of fighting not considered, this model provides the best case of a biologically textured game theory model with wide application.

We could say that the use of game theory has become routinised in biology. A biologist faced with a food-storing bird or spiders fighting over a web, or monkeys calling in the presence of predators, uses game theory in the same way an engineer uses mathematics in designing the pipes that cool a nuclear reactor. That is, without fanfare, without much interest in the formalities, but to get the job done effectively. The significant features of these studies are the memory of the food-storing bird and presence of vigilant competitors, whether the web is in a good site and how long it will last, how far the call is heard and how the receivers react. Only if there's a special kind of problem does the game theory analysis become prominent: usually, by contrast, the analysis looks like slightly sophisticated common sense. Modern empirical behavioural biology would be unrecognizably different without the major tool of game theory.

*Foundations of biological game theory*

I remarked earlier that game theory avoided one set of foundations in order to construct another, and also that my own current project is in the biological foundations of fitness optimization, in which game theory will play an important part. The two sets of foundations are thus being combined.

Game theory proper simply posits players, strategy sets, game structures and payoffs, but biology has its own more fundamental foundations—gene frequencies and their dynamics, organisms and their ecology and environment. To connect them is to link these two set of ideas, mending the very disjunction that I argued earlier permitted the initial flourishing of game theory in biology.

The connections have been discussed since at least Taylor and Jonker (1978) and are substantially discussed in Maynard Smith's 1982 book. His view was that game theory assumes the simplest possible genetics, and one chapter explores the consequences of more complicated genetics. There have been more systematic attempts to link optimization of fitness in various contexts to population genetics. The 'streetcar theory' of Hammerstein (1996) is the most advanced of these that specializes on game theory situations rather than simpler cases. My own 'formal Darwinism project' aims to make a very general link, but so far has not tackled the game theory case in which the effect of the actions of one individual on the fitness of itself or others may depend on the actions of others.

Why bother uniting these foundations? Sometimes there is doubt about what the payoff function should be when modeling a biological situation. Sometimes there is a doubt about whether a game is a reasonable biological game. Accepting offspring number as essentially the maximand imposes quite tight constraints on the set of reasonable games. One very general question, answered already by my project in the negative, is "Should organisms be risk-averse?" To be more exact, to obtain optimization results from population genetics, the maximand in the optimization must be individual fitness relative to the population mean. When we allow arbitrary uncertainty, the maximand is the probability-weighted arithmetic mean of that relative fitness. This allows a natural combining of maximization within and between generations.

The topic of bet-hedging in biology (see for example Seger and Brockmann 1987) is about risk aversion, but deals in absolute not relative fitness. The new result in no way contradicts the bet-hedging models, but allows a reinterpretation of them. With a colleague, Dr. Francisco òbeda de Torres, I am currently working on justifying the new foundations by arguing that the reinterpretation provides a better, more coherent, more biologically meaningful, understanding of the classical behaviour of bet-hedging models.

## What is the proper role of game theory in relation to other disciplines?

I was very fortunate to be invited to a number of seminars at the Zentrum fur Interdisciplinäre Forschung in the University of Bielefeld, in the late 1970s and early 1980s, organized by Reinhard Selten. I was a young graduate student, and while walking in a group including Selten, I put my view that what biologists did with ESSs wasn't strictly speaking game theory, but should instead be considered a different area. Selten was firm to the contrary that game theory was a broad church, and should include the biological branch as well as mathematical and economic branches, and other formal strategic reasoning as it developed. It made no sense to have to rediscover ideas. This certainly suited the seminars, which were inter-disciplinary, and Selten's words were wise indeed. (It was not my only experience of a warm open-heartedness among game theorists in place of the cold calculation that might naively be expected. At one Bielefeld meeting, a biologist was making interventions with an air of "nobody understands me, everyone is against me"; and a distinguished game theorist passed a note round the table to him that said simply "I'm on your side".)

One role, therefore, is to maintain an understanding of strategic behaviour that both acts as a repository of ideas for other subjects, and assists in transmitting ideas between different subjects.

## How biology contributes to game theory

Biology has brought a kind of thuggish brutality to the refined intellectual world of game theory. Most biologists knew little of game theory proper, and developed their own version of strategic thinking, which was very down-to-earth. For example, the distinction between player and automaton is not one that most biologists would even recognize: biologists routinely think of biological game theory as games being played between automata. Remarkably, this cloddish approach has succeeded in making contributions back to mathematical game theory, chiefly because hyper-rationality and equilibrium selection have come to seem too ethereal. Nowadays, evolutionary game theory (see Young 1998 for a substantial and sustained application) borrows ideas from biology to develop models and equilibrium concepts.

The spectrum between completely unfettered rationality and simple rules of thumb lies at the heart of an ongoing dilemma in game theory. The immediate paradox of abandoning unfettered rationality is that there are very many ways to do so, and it is hard to choose, and harder to justify, any one particular way. Biology has offered some ideas on less arbitrary ways, and it is interesting to consider where those ideas come from.

The first special property of biological applications is that natural selection does the optimizing, or at least does the hard work in bringing about the optimizing. If the individual organism is set up to maximize fitness, this is because it has been constructed with parameters tuned by natural selection to appropriate values for the environment it finds itself in. (And it is crucial in this regard whether on a proper understanding the action of natural selection is indeed to bring about maximizing behaviour—which is why my current research project is focused on that question.) There are two implications: first, if there are constraints that fetter rationality, there is still a maximizing process at the core; whereas in principle, once an economist abandons full rationality, there are many non-rational, non-optimising possibilities for how behaviour is organized. Second, if a question arises about what the constraints are, a biologist is happy to say "I'll investigate them as they arise", with confidence that experiment could in principle discover and specify informational constraints and even computational constraints. Even though this is rarely done, the biologist does not feel it is unacceptable to have apparently arbitrary constraints.

*How game theory contributes to biology*

Concepts such as individual advantage, fitness, strategic behaviour, were all present in biology before game theory arrived. It might be thought that game theory merely allows biologists to clarify the ideas they were already having, but this is to underestimate hugely the importance of formality. Constructing a formal framework takes powers of analysis and great imagination. The Nash equilibrium concept itself is, in one way, a very 'so-what', 'that's kind of obvious' idea. Yet to capture one obvious idea in a formal way can unleash further vast empires of ideas. Biology's central game concept, the Evolutionarily Stable Strategy, had its effect precisely by showing how to capture an idea in mathematics, that the individual should be the agent, that fitness or some fitness-proxy should be the payoff, and that the set of strategies should be the set of things an individual should do. Obvious, afterwards.

The formal structure extends the range of people who can contribute to a given topic. Until game theory arrived in biology, only those who studied animal conflict in the field could really think about it to any purpose. By building a bridge towards mathematicians and mathematically minded biologists, Maynard Smith and Price (1973) allowed them to have productive thoughts about animal conflict. The huge success of extending the range can be measured by observing that those who now study animal conflict in the field equip themselves with many theoretical ideas and expectations, and observing the enormous progress the field has made in thirty-five years.

There was an initial inflow of ideas, as already discussed, and many of those applying game theory ideas in biology today have benefited from textbooks and courses on game theory.

*Foundations of games / humans as optimizers*

There are two ways in which biology's relationship to game theory is special. Although I cite no one in this section, I make no claim to originality—these ideas are commonplaces in many circles, and are surely already written down. The first special feature is that biology provides good reasons to believe that some of the assumptions of game theory are upheld. Natural selection leads to organisms acting as optimizers, without any need for conscious calculation. Further, by selecting the appropriate behaviour, natural selection in effect builds in information, so that an organism acquires a prior probability distribution over states of nature. Thus, natural selection makes (the appearance of) rationality and the possession of necessary information entirely plausible. From this point of view, biology should be the envy of other fields applying game theory.

The second aspect is less formal but conceptually deeper. Suppose humans do play games well, and maximize some payoff function. Why do they do that? How did they acquire the skills? The biologist's answer is that humans are organisms subject to natural selection, and any complex ability they possess must, at some level, be attributed to natural selection.

Consider two levels. At the lowest, humans are good at catching balls. The hand-eye coordination required is uncontroversially a product of natural selection, even if the circumstances in which the skill is now mainly used are culturally constructed and inflected.

At a higher level, and sticking with games, humans consciously pursue strategies to attain goals, for example in chess and in football. That conscious maximizing ability is likely to be the product of natural selection, as a means to bring about adaptive behaviour in as flexible a way as possible. If natural selection builds an optimizing machine, and then manages to provide an appropriate maximand for that machine, then the organism will succeed in a much wider array of circumstances than if every response had to be separately evolved, and hard-wired. Thus, the optimizing and game-playing of humans is likely to be itself a product of natural selection.

Biology thus stands both as a special case in which the assumptions are more likely to be met than in other fields applying game theory, and as the discipline capable of explaining why humans play games in the first place. It is tempting to go a step further and mention that the raw materials for the conceptual and mathematical abilities used in the academic study of game theory must also be attributed to natural selection.

**What do you consider the most neglected topics and/or contributions in late 20th century game theory?**

**What are the most important open problems in game theory and what are the prospects for progress?**

Two issues strike me, but as game theory is such a large subject, I do not know whether these problems are still open. I can say only that their closure has not yet reached me.

The major issue concerns the use of information. In discussing the Sequential Assessment Game, I mentioned earlier that an arbitrary stipulation had to be made about how information was used, in order to prevent any signaling use of the information available. There are of course many signaling models in game theory. But signaling games get so complicated so quickly that it is hard to see how many realistic situations could be analysed. There is, then, a need for some perhaps heroic simplification or abstraction or, barely conceivably, perhaps just some mathematical heavy-lifting, to allow the analysis of games with both simple payoff effects and multiple exchanges of information. A colleague and I made a biologically-inspired suggestion some time ago, but it has not been developed (Grafen and Johnstone 1993). It would be disappointing if game theory had little to offer in these more realistic situations.

Finally, I have never liked the Iterated Prisoners' Dilemma as a model of cooperation. It is too sharp. A good task for game theory would be to develop a more useful model of cooperation. That better model might one day entertain children out of school, and would deserve to inspire their future academic careers.

## References

Ayer, A. J. 1936. *Language, Truth, and Logic.* Gollancz, London.

Brick, O. 1999. A test of the sequential assessment game: the effect of increased cost of sampling. *Behavioral Ecology*, 10: 726-732.

Dawkins, R. 1976. *The Selfish Gene.* Oxford University Press.

Dugatkin, L.A. and Reeve, H.K. 1998. *Game Theory and Animal Behavior.* Oxford University Press.

Enquist, M. and Leimar, O. 1983. Evolution of fighting behaviour: decision rules and assessment of relative strength. *Journal of Theoretical Biology*, 102: 387–410.

Enquist, M. and Leimar, O. 1987. Evolution of fighting behaviour: the effect of variation in resource value. *Journal of Theoretical Biology*, 127: 187–205.

Fisher, R.A. 1930. *The Genetical Theory of Natural Selection*, Oxford University Press.

Grafen, A. 1990. Biological signals as handicaps. *Journal of Theoretical Biology*, 144: 517–546.

Grafen, A. 1990. Sexual Selection Unhandicapped by the Fisher Process. *Journal of Theoretical Biology*, 144: 475–518.

Grafen, A. and Johnstone. R.A. 1993. Why we need ESS signalling theory. *Philosophical Transactions of the Royal Society* (London) B 340: 245–250.

Grafen, A. 1998. Fertility and Labour Supply in Femina economica. *Journal of Theoretical Biology*, 194: 429–455.

Hammerstein, P. 1996. Darwinian Adaptation, Population-Genetics and the Streetcar Theory of Evolution. *Journal of Mathematical Biology*, 34: 511–532.

Leimar, O. 1988. *Evolutionary analysis of fighting behaviour*. Ph.D.-dissertation, Stockholm University.

Lewontin, R. 1961. Evolution and the theory of games. *Journal of Theoretical Biology*, 1: 382–403.

Maynard Smith, J. and G.R. Price 1973. The Logic of Animal Conflict, *Nature*, 246: 15–18.

Riechert, S. E. 1998. Game theory and animal contests. In L.A. Dugatkin & H. K. Reeve eds. *Game Theory and Animal Behavior*. Oxford University Press. pp 64–92.

Seger, J. and Brockmann, H.J. 1987. What is bet-hedging? *Oxford Surveys in Evolutionary Biology*, 4: 182–211.

Selten, R. 1980. A note on evolutionarily stable strategies in asymmetric conflicts. *Journal of Theoretical Biology*, 84: 93–101.

Taylor, P.D. and Jonkers, L.B. 1978. Evolutionarily Stable Strategies and Game Dynamics. *Mathematical Biosciences*, 40: 145–156.

von Neumann, J. and Morgenstern, O. 1944. *Theory of Games and Economic Behavior*. Princeton University Press.

Young, H. P. 2001. *Individual Strategy and Social Structure* (Second Edition). Princeton University Press.

Zahavi, A. 1975. Mate selection – a selection for a handicap. *Journal of Theoretical Biology*, 53: 205–214.

Zahavi, A. 1977. The cost of honesty (Further remarks on the handicap principle). *Journal of Theoretical Biology* 67: 603–605.

Zahavi, A. and Zahavi, A. 1997. *The handicap principle: a missing piece of Darwin's puzzle.* Oxford University Press.

# 8
# Peter Hammerstein

Professor of Organismic Evolution
Institute for Theoretical Biology
Humboldt University Berlin, Germany

---

Why were you initially drawn to game theory?

In 1976 I worked as a mathematician in the biology department of the University of Bielefeld. My job was to act as an advisor and teach statistics as well as mathematical modelling in population biology and animal behaviour. It was quite an experience to see that many researchers' headaches had little to do with mathematics and statistics and much to do with a lack of conceptual clarity in biology. As a young discipline developing at breath-taking speed, biology was plagued by fuzzy ideas. Advising these biologists required slipping into the role of a theoretician and helping ask the right biological questions instead of just helping with the maths—not an easy thing to do for a novice in the life sciences. After reading the seminal paper by John Maynard Smith and George Price (1973) on the logic of animal conflict, I immediately felt that the emerging evolutionary branch of game theory could contribute some much-needed conceptual clarity and empirical focus to the biology of behaviour, morphology, and physiology. I thus started working on a biological Ph.D.-thesis in evolutionary game theory, initially without a supervisor and spurred by the critical comments of colleagues who considered this a heretic endeavour. At the *Fifteenth International Ethological Conference* in 1977, the eminent evolutionary biologist Mary Jane West-Eberhard strongly encouraged me to remain a heretic. She also helped me conquer my inhibition to approach celebrities by dragging me over to meet John Maynard Smith—the beginning of a long-lasting scientific alliance with the inventor of evolutionary game theory.

A year later, the ethologist Klaus Immelmann informed his colleagues at the University of Bielefeld that an international meeting at the Centre for Interdisciplinary Research needed to be organized within only a few months. Otherwise an existing opulent fund would have to be returned to the tax payers—an inconceivable act. Almost any theme was fine, yet nobody wanted to do it on such short notice. I decided to turn a chore into an opportunity to bring together all of the scientists who were beginning to apply game theory to biology. Klaus Immelmann agreed to this plan and made me – still a Ph.D.-student – the main scientific organizer of the symposium "Evolution and the Theory of Games", November 1977. What a brave decision for a professor in the hierarchical world of German universities at the time! It left me, though, with the problem of attracting all the established pioneers of the field. This seemed like just another inconceivable act. However, the meeting was going to be one of the first ever on evolutionary game theory. So, fortunately, it was possible to persuade John Maynard Smith to come and this made it a lot easier to lure Jane Brockmann, Nicholas Davies, Richard Dawkins, Alan Grafen, John Krebs, Manfred Milinski, Geoffrey Parker, and many others to Bielefeld. The symposium demonstrated impressively that the new field of evolutionary game theory, then called ESS theory, had indeed laid the foundation for meaningful empirical studies on animal behaviour, morphology, and physiology.

During the preparation of the symposium, I had drawn Reinhard Selten's attention to evolutionary game theory. He attended the symposium as the main representative of classical game theory. John Maynard Smith and Reinhard Selten met for the first time and John's private comment was: "Reinhard is a very impressive man. He is the first game theorist who understands what I am doing. Normally they try to teach me that I have simply reinvented classical game theory." It was very much in my interest that these two great minds get along well. During the symposium I asked John to become my biological Ph.D.-supervisor and Reinhard to join him as my game-theoretic co-supervisor. Since then my work in evolutionary biology investigates conflict and cooperation within and among organisms, bridging the disciplines of my mentors, and extending the field of game theory to genetics, immunology, and biological studies of human sociality and cognition.

## What example(s) from your work (or the work of others) illustrates the use of game theory for foundational studies and/or applications?

In the early seventies common sense among behavioural biologists held that that in many cases animals defended their territories because they had already invested in them, whereas intruders had not. It was extremely difficult to explain to biologists that they committed the sunk cost fallacy (what behavioural biologists often refer to as the "Concorde" fallacy) when using this as an evolutionary – as opposed to cognitive – argument. Maynard Smith and Parker (1976) therefore developed a simple game showing that a contest between an 'owner' and an 'intruder' of equal fighting ability can be settled by an evolutionarily stable 'owner wins' convention – without assuming an owner bias in the reward for winning. This raised a number of questions: If the intruder is stronger than the owner, can David win against Goliath using only an 'owner wins' convention? How can one analyze evolutionary games that take several asymmetric aspects into consideration and have thousands of pure strategies even in simple cases? What is the appropriate notion of a strategy for such an evolutionary game?

I tried to answer these questions (Hammerstein 1981) and it turned out that, yes, weaker owners can, within limits, deter stronger intruders when both play a particular evolutionarily stable strategy in a game with discrete levels of escalation. Thus, ownership in a stronger sense can evolve, though with important caveats (e.g., Hammerstein & Parker 1981, Grafen 1987). The appropriate notion of strategy for an owner-intruder game seems to be what classical game theorists would have called a behaviour strategy. A behaviour strategy assigns actions to roles, possibly randomizing them for a given role. More generally, I conceive of a strategy as an agent's program to translate information into action. It does not matter who or what generated the strategy, and the information comprises what the agent knows about the present and the past, including knowledge about its own internal physiology. The action can be behaviour in the conventional sense, a developmental step during ontogeny (Hagen and Hammerstein 2005), or the secretion of a chemical, such as a toxin or hormone. If anything is characteristic of the strategy concept, it is its conditionality. Unfortunately, biologists have had a hard time learning this, as witnessed by the wide use of the term 'conditional strategy'.

I should also briefly touch upon an interesting technicality. Maynard Smith originally defined an ESS as a particular symmetric Nash equilibrium of a symmetric game. Reinhard Selten and I think that even asymmetric conflicts require a symmetric game as the basic model. The reason is that an inherited strategy may have to be played by an animal in any of the roles of the game, let it be the roles of a smaller owner and a larger intruder, or even male and female (Hammerstein & Selten 1994). To illustrate the latter point, think of the genetic program for milk producing glands. This genetic program can be found in the DNA of both males and females, i.e. in organisms with distinctly different hormonal states (the roles). In one state the program induces development of the glands, but not in the other. Under appropriate assumptions, however, the evolutionary game can be decomposed into asymmetric subgames, such as that between a young female and an old male. The evolutionarily stable strategy then requires playing Nash equilibria in these asymmetric subgames (Hammerstein 1981, Selten 1981). This is how asymmetric Nash equilibria emerge in the theory of within species interactions.

What did game theory teach us about animal contests? Well, to cut a long story short, it helped us understand the sad but important fact that animal fights ending without serious damage are often not as peaceful as they appear. In many cases one contestant would inflict serious injury on the other if the latter did not escape at some crucial point. Asymmetries play a major role in conventional conflict resolution. The best analogy is that of two car drivers at an intersection: The one coming from the right pushes the accelerator, risking a collision, and the other hits the break.

Around 1980 I met the American spider biologist Susan Riechert. She told me that her desert spiders (*Agelenopsis aperta*) seemed to have read one of my manuscripts on asymmetric contests. My response was "too bad, otherwise we could use them to test my model." What I meant, of course, was that we knew so many facts about animal territoriality before producing our models that we could hardly claim we predicted the use of asymmetries in the proper sense of the term 'prediction.' However, it made a lot of sense to test the model assumptions and see whether they predicted the facts for realistic reasons. It took us several years to achieve this goal. We learned the hard way that it is much easier to solve a game than to determine game-theoretic payoffs in the field and estimate the costs and benefits of fighting in terms of

Darwinian fitness (Hammerstein and Riechert 1988, Hammerstein & Selten 1994). Before we conducted our study, some biologists had expressed the pessimistic view that, in practice, it would be impossible to determine the relevant payoffs in a field study. I am glad they did not stop us from trying.

In Susan Riechert's major study area, spiders face a harsh environment in which web sites are in short supply. Web site tenancy crucially affects survival probability, fighting ability, and the rate at which a female spider produces eggs. Competition for sites causes a great number of territorial interactions. At an average web site ensuring a moderate food supply, a contest usually ends without an injurious fight. For small differences in weight, the owner typically wins, whereas a much heavier intruder will gain possession of the site. Susan and I entertained ourselves as spider real estate brokers and asked the question: How much would a site be worth to a spider? It turned out that the daily weight gain was about 5.1 mg for an average site and 7.9 mg for a 'posh' location. Cranking the field data through a model for spider life histories, we learned that the strategic values in terms of female fitness (gain minus opportunity cost) were 20 mg and 72 mg egg mass for the respective sites.

At first glance this may look like boring quantitative biology. Yet, these numbers reveal an interesting effect: Daily weight (and also energy) gains at a posh web site are only about 50 percent above that of an average site, but the value of winning is more than tripled. Why is there such a strong multiplier effect? In the spider 'society' that we studied, owners of posh sites gained weight at a rate which is near the maximum for the population under consideration. As a result, an individual's weight relative to the population mean will increase as long as it inhabits the excellent site. This in turn is important for remaining in possession of the site because weight matters in territorial defence. Therefore, an originally 'richer' spider will have better future chances of maintaining its wealth than an otherwise equal 'poorer' individual. This explains the multiplier effect. Anthropomorphically speaking, in Susan Riechert's best studied population in New Mexico the gap between rich and middle class spiders increases dramatically over time. Without the methodological discipline that game-theory imposed on us, we would have been blind to this phenomenon.

To conclude the spider example, we also calculated the negative consequences of fighting. Our macabre results included that loss of a leg costs a female 13 mg egg mass while death costs 102 mg. The main question was then: Do we now understand why it does not pay a member of this spider society to deviate from the contest behaviour typically observed? It was possible to show, in particular, that intruders would pay a high price for "breaking the rules" of this society by disregarding the status of an owner who is not much heavier. This cost is high enough to ensure the evolutionary stability of the "owner wins" convention. We had identified a Nash equilibrium in a field study.

I now draw your attention to a different subject: Evolutionary game theory and genetics. From a 'hard-nosed' population geneticist's point of view, the concepts of an evolutionarily stable strategy or Nash equilibrium are problematic because they overemphasize the adaptive power of natural selection. In sexually reproducing populations, phenotypic optimisation can be impeded by pleiotropic gene effects, recombination, epistasis, selfish genetic elements, etc. A 'hard nosed adaptationist' would counter this by saying that biologists have studied so many phenomena that turned out to be highly adaptive that population geneticists may sometimes underestimate the adaptive power of natural selection. Birds have wings with superb aerodynamic properties. These wings are not cast into quadratic shapes by genetic constraints, as Maynard Smith would have put it. Karlin (1975) and some of his followers expressed the contrasting view that selection models with several genes and recombination do not support the idea of fitness optimisation in even the simplest possible case where genotypes have constant fitness independent of genotype frequencies and population densities. In the same spirit Moran (1964) had advocated the "non-existence of adaptive topographies". All this bothered Maynard Smith and other evolutionary game theorists a bit. An emotional discourse ensued that made it difficult to mediate between phenotypic and genetic modellers of evolution.

But if the theory does not permit the bird to have its wings, there must be a problem with the theory. Mathematically the statements of Karlin and Moran were undisputable. The problem, therefore, must lie in the concepts used. In extending work by Eshel and others (Eshel and Feldman 1984, see also Hammerstein 1996b for more details about the history of the ideas involved), I tried to clarify the conceptual confusion that made it so difficult for many geneticists and many students of phenotypic evolution to appreciate each others' evolutionary insights (Hammerstein and Selten 1994, Hammerstein 1996a, Hammerstein 2005). This clarification is based on the idea that evolution will change the genetic system itself when genetic constraints become a strong impediment to phenotypic adaptation.

The simplest of all possible cases is that of a model with one locus, two alleles, and a heterozygote fitness advantage like in sickle cell anaemia. Ignoring genetics, a naive believer in adaptive phenotypic evolution would have to think that the phenotype associated with the heterozygote should increase in frequency from generation to generation until it goes to fixation. Of course, we know that Mendelian segregation will not allow the population to consist entirely of the optimal phenotype. However, the model we use to derive this trivial conclusion is quite narrow in scope. If our model included, for example, the possibility of gene duplication followed by a crossover event, this process could combine the two alleles of the heterozygote on the same chromosome. Once the genetic rearrangement has taken place, the population would lose the less adaptive phenotypes associated with the homozygotes of the original gene. Another way of overcoming something like sickle cell anaemia by genetic rearrangement is to think of a new mutation at some locus in the genome that has not yet been included in the mathematical model. Suppose that this new mutation has a dominant phenotypic effect that is equivalent to that of the heterozygote. Such a gene would spread and the sickle cell allele would decrease in frequency. The message is that if genetic constraints cause strong maladaptations, one can at least imagine ways of how these constraints could be resolved by evolutionary mechanisms.

To summarize this I invoke a journey on a streetcar (Hammerstein 1996a). The streetcar stands for an evolving population. Its passengers are meant to be the genes in this population. At least two loci that can undergo recombination are considered. Attention is drawn to the stops (equilibrium states of the evolving population). At each stop passengers (new alleles or duplicate genes) enter or leave the streetcar and within the streetcar recombination takes place. Imagine now that the streetcar stops and the phenotypes are not adapted to the environment. If one takes into account a particularly wide range of potential mutations, or of appropriate duplication and crossover events, then the streetcar will not remain at this maladaptive equilibrium. The streetcar starts moving again after an appropriate new passenger enters it that undermines the genetic constraint responsible for this particular maladaptive equilibrium. In this sense, phenotypically maladaptive equilibria are 'temporary stops' of the streetcar. Suppose now that the streetcar has reached a 'final stop' where the extension procedure just discussed will not cause a phenotypic change. A number of mathematical results, starting with a seminal paper by Eshel and Feldman (1984), suggest that a final stop is necessarily a phenotypically adaptive state—the bird has its wings (see also Hammerstein and Selten 1994; Eshel 1996; Hammerstein 1996a,b; Weissing 1996).

Why does the streetcar paradigm reconcile genetic and phenotypic approaches to evolution? Temporary stops depend on genetic detail, and this is the domain of population genetics theory. Final stops depend on phenotypic adaptation because genetic rearrangements have removed all genetic constraints. It is an empirical question whether the human observer of natural phenomena would see more streetcars at final than at temporary stops. Sickle cell anaemia, for example, is a clear case of a temporary stop. A philosophical comment is that one cannot hold the argument of genetic constraints against a theoretical research programme where the search is for final stops. Conversely, one cannot criticise an emphasis on maladaptive properties if the research programme is directed toward temporary stops. Neither genetic nor phenotypic modellers of evolution will be losing face if they subscribe to the streetcar philosophy.

A word must now be said about evolutionary game theory with underlying diploid genetics and more than one gene. For a large class of games a final stop corresponds necessarily to a Nash equilibrium. The final stop thus has a property that is central to decision theory and demonstrates that evolution can act like a rational decision maker (Hammerstein 1996a). For a long time, humans tended to claim a 'monopoly' on rational decision making when comparing themselves with the animal world. From recent literature in social psychology and experimental economics we know, however, that human decision processes deviate systematically from the principles of economic rationality. Paradoxically we learn from biology that animals often behave as if they had taken a course in rational economic decision making. The streetcar theory resolves this paradox and can be seen as the conceptual 'backbone' of evolutionary game theory in biology.

## What is the proper role of game theory in relation to other disciplines?

Mathematicians in game theory often think that their work provides tools ready to be used by members of other disciplines. True interdisciplinary exchange requires more than pulling tools off the shelf, however. The challenge is to readjust them or invent new ones. Some tools may need little adjustment, others turn out to be unadjustable. In the following I present a number of examples from biology.

**Example 1.** Robert Aumann invented a powerful game-theoretic concept named 'correlated equilibrium' (Aumann 1974). The idea was that players should make use of appropriately chosen correlated signals to decide what strategy profile to play. This made it possible to increase one's expected payoff beyond that of a simple Nash equilibrium. In Aumann's approach the crucial signals are not explicitly modelled. The rationale of this simplification would be that correlating signals are always amply available – no need to worry about that. In the human world this rationale may be convincing. When studying the animal world, however, the correlating random event should be modelled explicitly, since it is an important part of the description of the phenomenon under investigation, and biologists do actually worry about what signals are available (Hammerstein and Selten 1994). There is thus a subtle but important difference between (a) applying the correlated equilibrium concept to the Hawk-Dove game of animal contest behaviour and (b) studying the asymmetric version of this game with an explicitly modelled owner-intruder asymmetry. The example shows nevertheless how closely economic game theory relates to biological evolutionary game theory.

**Example 2.** An even better example to illustrate the close link between game theory and biology is Michael Spence's signalling theory. What he published long ago (Spence 1973), is essentially the same as Amoz Zahavi's handicap principle (Zahavi 1975). Spence's model of education as a credible signal of productivity (Spence 1973) is little known in biology. In his school days he must have had doubts – like many of us – about the value of learning 'exotic' things, such as Latin, or the capital cities of Europe. So he was intrigued by the question: Would education have a positive effect on wages if it did not at all increase a person's productivity? In Spence's signalling game, a person chooses a level of education conditional upon talent. Education is not free, and it is less costly for the more talented to continue in school. Education level is the signal that firms observe when hiring their personnel. Spence found that a game-theoretic equilibrium can exist in which education signals talent and higher education implies higher wage. The reason is that only the talented persist in long years of schooling, so employers use years of schooling as a signal of the unobservable trait, talent.

Education plays the same role in Spence's model as the peacock's tail in Zahavi's handicap principle. It took biologists a long time to accept Zahavi's idea and they created signalling models very similar to that of Spence in their own domain (Pomiankowski 1987, Grafen 1990). It would have been so much easier for biologists to simply adjust the conceptual tool that Spence had created before Zahavi started advocating his handicap principle.

**Example 3.** After highlighting some of the interesting similarities between game theory in economics and biology, I now switch to important dissimilarities that can be illustrated through biological markets. Ever since Darwin read the classical economist Thomas Malthus, the emergent properties of competitive interactions have been prominent in biological thinking. The analogy between animal mating and human trade led much later to the metaphor of mating markets. Recently a more general field of research on biological markets has emerged (Noë and Hammerstein 1994, 1995, Schwartz and Hoeksema 1998, Simms and Taylor 2002, Bshary and Noë 2003, Bowles and Hammerstein 2003). As Ronald Noë and I have argued, the reason behind this broadening of scope is that partner choice plays an important role in social interactions other than mating and that many cooperative exchanges take place within and between species.

At first sight, research on markets provides many opportunities for fruitful interdisciplinary discourse with economists. The relationship between cleaner fish and their customers, for example, nicely demonstrates the economic principle of monopolistic competition: Buyers with few alternative sources of supply will have less advantageous transactions than will those who can shop around. Cleaners live in coral reefs and have customers from the immediate neighborhood and the open sea. Those from the immediate neighbourhood would risk being eaten by a shark or whatever predatory fish if they travelled long distance. Redouan Bshari discovered in his empirical studies that local customers, for whom long-distance moves are costly, are cleaned less well than are long-range travellers, who can exert partner choice (see Bshari and Noë 2003 for a review).

Driven by enthusiasm for economic thinking, biologists – including myself and my long-standing colleague Ronald Noë – have frequently used the market metaphor without probing deeply enough into what its use entails (Bowles and Hammerstein 2003). The famous 'mating market', for example, seemed to be a straightforward application of ideas borrowed from economics. This is a two-sided market in which one side offers eggs to be fertilized and the other sperm to fertilize them. What determines the number of sperm and the number of eggs that are produced? A neoclassical economist would probably want to apply the conventional Walrasian market model. This model invokes the law of supply and demand, which states that, in a market economy, supply and demand should equalize over time, a process called 'market clearing'. Is there market clearing in the mating market?

In many species, the supply of sperm in a population is significantly larger than that needed to fertilize all the eggs. The law of supply and demand therefore does not apply, but why not? First, a disciple of neoclassical economics would expect individuals to prefer entering the market on the side with the scarce good, here, eggs. Yet, males and females appear in approximately equal numbers in most species. If Karl Düsing (1883) and Ronald Fisher (1930) had not already created sex ratio theory, the failure of our economist's expectation would have led biologists to develop it. Second, our economist would suggest that, when supply exceeds demand, a female would 'sell' her eggs for sperm plus additional 'commodities', such as a nuptial gift or paternal care. However, only some species offer such commodities. The reason is that, unlike trade in contemporary human market economies, *animal trade is not subject to enforceable contracts*. There are no police to arrest males that fail to pay what they promised for the eggs that they fertilize. Contrary to the expectation of 'traditional' neoclassical economics, biological markets will frequently not clear (Bowles and Hammerstein 2003).

The same puzzle arises in the theory of human labour markets. If labour is chronically in excess to supply, what prevents the unemployed workers from offering employers a more attractive package, promising to work harder for the same wage? Or given that markets do not clear, so that jobs are typically scarce and workers abundant, why do employers not sell jobs, charging a fee to the prospective worker as a condition of employment? Needless to say, Economists are now realizing that many human markets also do not clear. Josef Stiglitz even talks of "the abrogation of the law of supply and demand". One of the reasons is again an *inability to enforce contracts*. In traditional economic models, this was not apparent because prices were determined without explicitly representing the interactions among traders. Unlike the typical biological modelling style, no account was given of who meets whom, what the traders know, and how they settle on a transaction. I am tempted to say that this traditional approach was lacking the key ingredients of strategic analysis if one understands the term strategy the way I described it earlier in this interview.

The new, post-Walrasian market theory, however, is different. Its assumptions include the incomplete nature of contracts (what biologists would refer to as the potential for cheating) and the traders' limited information about the trades being offered and accepted by other traders. In recent post-Walrasian models of labour markets, credit markets and markets for goods of variable quality, market clearing does not occur (Stiglitz 1987, Bowles 2003), even in a competitive equilibrium. Here, we see a new convergence between biological and economic modelling approaches. There are, thus, good prospects for joint brainstorming in the "second wave of evolutionary economics in biology" that I proclaimed together with my colleague Edward Hagen (Hagen and Hammerstein 2005).

**Example 4.** Cooperative game theory appears to me to be the problematic tool in the toolbox. This branch of game theory centres around the assumption of Pareto optimality. It thus assumes cooperation and then tries to understand the details of it. In contrast, the reason we have introduced game theory in biology is to understand the strong limitations evolution seems to have imposed on cooperation. Wherever strong forms of cooperation have emerged, we surely want to explain them. Neither the biological facts nor our own theory have pushed us biologists much in the direction of cooperative games (Parker and Hammerstein 1985).

## What do you consider the most neglected topics and/or contributions in late 20th century game theory?

I recently asked my students what they think about an assumption widely used in conventional game theory, namely that "everybody knows that everybody knows that everybody in the game is rational". One of them responded quickly, "this assumption is great, it would allow a fictitious mathematician from Mars to theorize about us without ever visiting the earth". My response was "whom do you mean by *us*?"

If game theory committed any serious sin in the $20^{th}$ century, it certainly was the *negligence of the facts* that was caused by its long-lasting attachment to the super-rational – I am tempted to say super-natural – homo oeconomicus. But even when the monument of homo oeconomicus started to crumble, some attempts to put a new statue on its pedestal came close to a remake. What sense does it make, for example, to acknowledge the boundedness of rationality by assuming limited memory and then let the same player have a superbrain with unlimited calculation power, solving any optimization problem at no cost in no time?

Once economics had started paying serious attention to Maynard Smith's evolutionary game theory, another problem arose. In economics one can, for example, conceive of the replicator dynamics – our simplest model for natural selection – as a social learning process (Hofbauer and Sigmund 1998), and then do 'Maynard Smithian economics'. Learning would to some extent produce 'quasi rational' behaviour. The Martian mathematician, if he knew that this is how we learn, could then theorize about us without ever moving his spacecraft. The problem is, of course, that real learning in humans does not necessarily resemble the replicator dynamics. Many other learning processes are conceivable. Reinhard Selten and I argued long ago that even evolutionary theory cannot help us too much with predicting which learning procedures one would expect because it seems impossible to delineate appropriate optimization models (Selten and Hammerstein 1984).

To illustrate how strongly human learning can differ from anything similar to the replicator dynamics, consider the well known phenomenon called 'the winner's curse'. This phenomenon was found in many real-life auctions of oil fields, broadcasting frequencies, etc. Essentially, what happens is that businessmen who overestimated the value of an oil field most were the ones who received it in the end. Not surprisingly, many of them ended up with an unprofitable deal. Selten (1975) did a repeated auction experiment in which the players had a fair chance to learn avoiding the winner's curse. They did not. They would have learned it, however, had their learning any resemblance with replicator dynamics (Hammerstein 2001). Well, I am inviting the Martian to land on earth.

Here is another problem. It is important in theoretical biology as well as in game theory to construct so-called toy models, let it be the game of Chicken or variants of the prisoner's dilemma. Toy models are wonderful for learning to what one should be sensitive and what pitfalls should be avoided in strategic analysis. It would be difficult to imagine any progress in game theory without thinking of a professional *homo ludens* playing with his or her toys. So, what is the problem? While children quickly get bored with their toys, we sometimes keep ours a little too serious. Repeated games may serve as an example. It is an eye opening experience to discover the logic of cooperation in repeated games; no wonder many of us have been talking about it over and over again. However, the mental thrill of studying the so-called folk theorems does not tell us anything about nature. It only means that the wisdom of the folk theorems will spread quickly within the scientific community and beyond – very much like a successful meme in the sense of Richard Dawkins.

Biologists have been trying for decades to come up with convincing animal examples. The evidence is surprisingly scarce. In my view there are a number of reasons (Hammerstein 2003). Many biological games are not well-described as repeated games. Where repeated games exist, they are often played by genetically related individuals so that kin selection is operating. Finally, and this may be the biggest problem, it is easy to underestimate the cognitive skills needed. Think of 'reciprocal altruism'. In real life the mental machinery has to perform complex cognitive tasks in order to achieve reciprocity based on partner control. If the partner, for example, fails to exhibit a cooperative act, this poses the attribution problem to determine whether the observed behaviour really belongs to the class of non-cooperative moves. The mental updating machinery must solve this problem. Routine learning may interfere with its information processing, which may be costly and error prone. To neglect all this is, of course, the secret why game theory is so beautiful.

## What are the most important open problems in game theory and what are the prospects for progress?

Biology has many open problems waiting for strategic analysis. Many parasitic nematode worms, for example, seem to 'know' extremely well how to play 'games' against the immune systems of their mammalian hosts (Hammerstein 2005). Once inside a host, the worms are continuously exposed to an array of dangerous effector mechanisms of the immune system but survive for very long time periods. How do they evade the immune response? That is the key question. Nematodes secrete cystatins that inhibit, for example, proteases involved in antigen processing and presentation, leading to a reduction of T cell responses in the host. Cystatins of parasitic nematodes also modulate host cytokine response in a way that inhibits the Th1 response, thereby creating an anti-inflammatory environment (Hartmann and Lucius 2003). In essence this means that the worms are manipulating the immune system's internal communication to their own advantage. Worms teach us something about 'weak spots' in the immune system's signalling networks.

Note that the nematode game takes place at the molecular level. A parasite strategy involves releasing molecules that interfere with molecular networks. Conversely, the immune system could in principle rearrange its communication networks in order to be safe. One would, therefore, expect that the immune system's complex organisation reflects long-term exposure to pathogens over evolutionary time. The issue is enormously complicated by the fact that immunity is always challenged by many different and ever-changing pathogens. The nematode under consideration is only one of them. Therefore, in a theoretical analysis, any rearrangement of the cytokine network needs to be judged by how it affects the immune system's overall performance in the evolutionary game with *many different opponents*. The network needs to be *robust* (Hammerstein et al. 2006) against a variety of possible attempts by pathogens to modulate the system to their advantage. What are the prospects for progress in this research on molecular warfare? It will be very difficult but I am sure the nematode worms will teach us something interesting about the *evolved functionality of the immune system* – a hot topic in evolutionary medicine.

There are many more items with a distinct game-theoretic flavour on the biologist's list of open problems. They relate, for example, to phenotypic plasticity (Hagen and Hammerstein 2005, Leimar et al. 2006), purification of mixed strategies (Leimar et al. 2004), intragenomic conflict and cooperation (Hammerstein & Hagen 2006, Hammerstein & Leimar 2006), endosymbiont theory (Engelstädter et al. 2007, Flor et al. 2007, Telschow et al 2002, 2005), indirect reciprocity (Leimar and Hammerstein 2001), and even to the origin of music (Hagen and Hammerstein, in press).

It is impossible to think about games involving the immune system without taking its mechanistic aspects and structural organisation into account. If we knew enough about the brain we would probably make a similar statement regarding the games studied in behavioural economics. Even though we still know little about how the brain makes complex decisions at the neural level, we can hardly avoid making assumptions about *cognition* whenever we interpret experimental results. The recent literature on altruistic punishment illustrates this impressively. An intriguing observation in experiments with anonymously interacting subjects is that punishment occurs even if it is costly and of no advantage to the punisher. One can have very different intuitions about how to interpret this. One intuition is that the assumption of pure self-interest is violated, suggesting a key role for genetic or cultural *group selection*. It seems a little odd, though, that subjects would *frame* the situation cognitively as an in-group encounter, since in the anonymous situation they do not even know what social group the other players are from. Another intuition is based on mismatch: subjects irrationally assume non-anonymity. But many experiments show that people are capable of understanding anonymity. All this indicates that experimenters will have to invest a lot in exploring how subjects cognitively frame the artificial scenarios in the laboratory, and how they actually process information (Hagen and Hammerstein 2006).

Emotions have often spurred the lively discourse of game theory. In the $21^{st}$ century we have to think *about* them. Anger, for example, plays an important role in punishment. It is conceivable that the 'switch' for this emotion is blind to anonymity, whereas we are perfectly capable of reasoning about it. This would perhaps solve the puzzle of altruistic punishment found in the laboratory. There are new puzzles, however. We normally think that players in a game should make the best use of all the information they have. But the brain does not seem to work like that. Some of the information that is available within the brain will always be kept away from decision processes. This is one of the secrets why the brain works so well. We must take this into account if we want to enrich game theory with empirical content.

A final word must be said about the overlap of core issues in biology and economics. The two disciplines have exchanged many concepts over the last decades. Economists' renewed interest in the detailed empirical properties of individual agents, however, brought their style of theorizing closer to that of biology and represents an important step towards a unified theory of animal and human behaviour. Edward Hagen and I call this the second wave of evolutionary economics in biology (Hammerstein & Hagen 2005).

# References

Aumann, R.J. (1974). Subjectivity and correlation in randomized strategies. *Journal of Mathematical Economics*, 1, 67–96.

Bowles, S. (2003). *Microeconomics: Behavior, Institutions, and Evolution*. Princeton, NJ: Princeton University Press.

Bowles, S. & Hammerstein, P. (2003). Does market theory apply to biology? In *Genetic and Cultural Evolution of Cooperation*, ed. P. Hammerstein, pp. 153–165, Cambridge, MA: MIT Press.

Bshary, R. and Noë, R. (2003). Biological markets: the ubiquitous influence of partner choice on the dynamics of cleaner fish – client reef fish interactions. In *Genetic and Cultural Evolution of Cooperation*, ed. P. Hammerstein, pp. 167-184, Cambridge, MA: MIT Press.

Engelstädter, J., Hammerstein, P. & Hurst, G.D.D. (2007). The evolution of endosymbiont density in doubly infected hosts. *Journal of Evolutionary Biology*, 20, 685–695.

Düsing, K. (1883). Die Factoren welche die Sexualität entscheiden. *Jenaische Zeitschrift für Naturwissenschaft*, 16, 428–464.

Eshel, I. (1996). On the changing concept of population stability as a reflection of a changing point of view in the quantitative theory of evolution. *Journal of Mathematical Biology*, 34, 485–510.

Eshel, I. & M.W. Feldman (1984). Initial increase of new mutants and some continuity properties of ESS in two locus systems. *American Naturalist*, 124, 631–640.

Fisher, R. (1930). *The Genetical Theory of Natural Selection*. Oxford: Clarendon.

Flor, M., Hammerstein, P. & Telschow, A. (2007). Wolbachia-induced unidirectional cytoplasmic incompatibility and the stability of infection polymorphism in parapatric host populations. *Journal of Evolutionary Biology*, 20, 696–706.

Friedman, J.E. & Hammerstein, P. (1991). To trade, or not to trade; that is the question. In *Game Equilibrium Models I: Evolution and Game Dynamics*, ed. R. Selten, pp. 257–275. Berlin: Springer Verlag.

Grafen, A. (1987). The logic of devisively asymmetric contests: respect for ownership and the desperado effect. *Animal Behaviour*, 35, 462.

Hagen, E.H. & Hammerstein, P. (2005). Evolutionary biology and the strategic view of ontogeny: Robustness versus flexibility in the life course. *Research in Human Development*, 2, 87–101.

Hagen, E.H. & Hammerstein, P. (2006). Game theory and human evolution: A critique of some recent interpretations of experimental games. *Theoretical Population Biology*, 69, 339–348.

Hagen, E.H. & Hammerstein, P. (in press). Did Neanderthals and other early humans sing? Insights from the territorial advertisements of primates, lions, hyenas, and wolves. *Musicae Scientiae*.

Hammerstein, P. (1981). The role of asymmetries in animal contests. *Animal Behaviour*, 29, 193–205.

Hammerstein, P. (1984). The biological counterpart to non-cooperative game theory. *Nieuw Archief voor Wiskunde (4)*, 2, 137–149.

Hammerstein, P. (1989). Biological Games. *European Economic Review*, 33, 635–644.

Hammerstein, P. (1995). Evolutionary biology: A twofold tragedy unfolds. *Nature*, 377–478.

Hammerstein, P. (1996a). Darwinian adaptation, population genetics and the streetcar theory of evolution. *Journal of Mathematical Biology*, 34, 511–532.

Hammerstein, P. (1996b). Streetcar theory and long-term evolution. *Science*, 273, 1032.

Hammerstein, P. (2001). Economic behaviour in humans and other animals. In *Economics in Nature*, eds. R. Noë, J.A.R.A.M. van Hooff & P. Hammerstein, 1–19. Cambridge: Cambridge University Press.

Hammerstein, P. (2001). Evolutionary adaptation and the economic concept of bounded rationality – a dialogue. In *Bounded Rationality: The Adaptive Toolbox*, eds. G. Gigerenzer & R. Selten, pp. 71–81. MIT Press.

Hammerstein, P., ed. (2003). *Genetic and Cultural Evolution of Cooperation*. Cambridge, MA: MIT Press.

Hammerstein, P. (2003). Why is reciprocity so rare in social animals? A protestant appeal. In *Genetic and Cultural Evolution of Cooperation*, ed. P. Hammerstein, pp. 84–93, Cambridge, MA: MIT Press.

Hammerstein, P. (2005). Strategic analysis in evolutionary genetics and the theory of games. *Journal of Genetics*, 84, 7–12.

Hammerstein, P. & Hagen, E.H. (2005). The second wave of evolutionary economics in biology. *Trends in Ecology and Evolution*, 20, 604–609.

Hammerstein, P. & Hagen, E.H. (2006). Broken cogs or strategic agents? *Science*, 304, 964.

Hammerstein, P., Hagen, E.H., Herz, A.V.M. & Herzel, H. (2006). Robustness: A key to evolutionary design. *Biological Theory*, 1, 90–93.

Hammerstein, P. & Hoekstra, R.F. (1995). Evolutionary theory: Mutualism on the move. *Nature, London*, 376, 121–122.

Hammerstein, P., Laubichler, M. & Hagen, E.H. (2006). The strategic view of biological agents. *Biological Theory*, 191–194.

Hammerstein, P. & Leimar, O. (2006). Cooperating for direct fitness benefits. *Journal of Evolutionary Biology*, 19, 1400–1402.

Hammerstein, P. & Parker, G.A. (1982). The asymmetric war of attrition. *Journal of Theoretical Biology*, 96, 647–682.

Hammerstein, P. & Parker, G.A. (1987). Sexual selection: Games between the sexes. In *Sexual Selection: Testing the Alternatives*, eds. J.W. Bradbury & M.B. Andersson, pp. 119–142. Dahlem Konferenzen. Chichester: John Wiley & Sons.

Hammerstein, P. & Riechert, S.E. (1988). Payoffs and strategies in territorial contests: ESS analyses of two ecotypes of the spider *Agelenopsis aperta*. *Evolutionary Ecology*, 2, 115–138.

Hammerstein, P. & Selten, R. (1994). Game theory and evolutionary biology. In *Handbook of Game Theory with Economic Applications, Vol.2*, eds. R.J. Aumann & S. Hart, pp. 929–993. Amsterdam: Elsevier.

Hartmann, S. & Lucius, R. (2003). Modulation of host immune responses by nematode cystatins. *International Journal of Parasitology*, 33, 1291–1302.

Karlin, S. (1975). General two-locus selection models: some objectives, results and interpretations. *Theoretical Population Biology*, 7, 364–398.

Kuhn, H.W., Harsanyi, J.C., Selten, R., Weibull, J.W., van Damme, E., Nash, J.F. & Hammerstein, P. (1995). The work of John F.

Nash Jr. in game theory: Nobel Seminar, December 8, 1994. *Duke Journal of Mathematics*, 81, 1–29. Also published 1996 in the *Journal of Economic Theory*, 69, 153–185.

Laubichler, M.D., Hagen, E.H. & Hammerstein, P. (2005). The strategy concept and John Maynard Smith's influence on theoretical biology. *Biology and Philosophy*, 20, 1041–1050.

Leimar, O. & Hammerstein, P. (2001). Evolution of cooperation through indirect reciprocity. *Proceedings of the Royal Society, B*, 268, 745–753.

Leimar, O. & Hammerstein, P. (2006). Facing the facts. *Journal of Evolutionary Biology*, 19, 1403–1405.

Leimar, O., Hammerstein, P. & Van Dooren, T.J.M. (2006). A new perspective on developmental plasticity and the principles of adaptive morph determination. *American Naturalist*, 167, 367–376.

Leimar,O., van Dooren, T. J.M. & Hammerstein, P. (2004). Adaptation and constraint in the evolution of environmental sex determination, *Journal of Theoretical Biology*, 227, 561–570.

Maynard Smith, J. & Parker, G.A. (1976). The logic of asymmetric contests. *Animal Behaviour*, 24, 159–175.

Maynard Smith J & Price GR (1973). The logic of animal conflict. *Nature*, 246, 15–18.

McElreath, R., Clutton-Brock, T.H., Fehr, E., Fessler, D.M.T., Hagen, E.H., Hammerstein, P., Kosfeld, M., Milinski, M., Silk, J.B., Tooby, J. & Wilson, M.I. (2003). Group report: The role of cognition and emotion in cooperation. In *Genetic and Cultural Evolution of Cooperation*, ed. P. Hammerstein, pp. 125–152, Cambridge, MA: MIT Press.

Moran, P.A.P. (1964). On the nonexistence of adaptive topographies. *Annals of Human Genetics*, 27, 383–393.

Noë, R. & Hammerstein, P. (1994). Biological markets: Supply and demand determine the effect of partner choice in cooperation, mutualism and mating. *Behavioural Ecology and Sociobiology*, 35, 1–11.

Noë, R. & Hammerstein, P. (1995). Biological markets. *Trends in Ecology and Evolution*, 10, 336–339.

Noë, R., van Hooff, J.A.R.A.M. & Hammerstein, P., eds. (2001). *Economics in Nature*. Cambridge: Cambridge University Press.

Parker, G.A. & Hammerstein, P. (1985). Game theory and animal behaviour. In *Evolution – Essays in Honour of John Maynard Smith*, eds. P.J. Greenwood, P.H. Harvey & M. Slatkin, pp. 73–94. Cambridge: Cambridge University Press.

Pomiankowski, A. (1987). The "handicap principle" does work—sometimes. *Proceedings of the Royal Society, B*, 127, 123–145.

Riechert, S.E. & Hammerstein, P. (1983). Game theory in the ecological context. *Annual Review of Ecology and Systematics*, 14, 377–409.

Schwartz, M.W. & Hoeksema, J.D. (1998). Specialization and resource trade: Biological markets as a model of mutualisms. *Ecology* 79, 1029–1038.

Selten, R. & Hammerstein, P. (1984). Gaps in Harley's argument on evolutionarily stable learning rules and in the logic of "tit for tat". *The Behavioral and Brain Sciences*, 7, 115–116.

Simms, E.L. & Taylor, D.L. (2002). Partner choice in nitrogen-fixation mutualisms of legumes and rhizobia. *Integrative and Comparative Biology*, 42, 369–380.

Spence, M. (1973). Job market signalling. *Quarterly Journal of Economics*, 87, 355–374.

Stiglitz, J.E. (1987). Causes and consequences of dependence of quantity upon price. *Journal of Economic Literature*, 25, 1–48.

Weissing F. J. (1996). Genetic versus phenotypic models of selection: can genetics be neglected in a long-term perspective? *Journal of Mathematical Biology*, 34, 533–555.

Zahavi, A. (1975). Mate selection – a selection for a handicap, *Journal of Theoretical Biology*, 53, 205–214.

# 9

# Sergiu Hart

**Kusiel and Vorreuter University Professor**

Institute of Mathematics, Department of Economics, and
Center for the Study of Rationality
The Hebrew University of Jerusalem, Israel

---

*Interviewed at the 17th International Conference on Game Theory at Stony Brook University, July 13, 2006, by Pelle Guldborg Hansen.*

Q: Professor Hart, why were you initially drawn to game theory?

A: The short answer is "Aumann." Bob Aumann came to Tel Aviv University to give a course on game theory when I was a second-year math student there. At the time I didn't know anything about game theory, but it sounded interesting. I took the course, and became quite excited. What I found so exciting about game theory, and still do, is that it is such a varied topic—in terms of questions, answers, methodologies, and so on. In mathematics one specializes in one area, say probability theory, and then does probability theory. But in game theory one can do probability theory one day and combinatorics the next, as well as logic, computer science, and biology. Let's just say there is never a dull moment. It's very varied, and for me that was extremely attractive. Of course, so was Aumann's personality. He knows how to fire people up. So that's how I started, and I don't regret it.

Q: What was your first impression of Aumann? Did you know that he was going to lure you into game theory?

A: No, I was just a student in his class. Then in my third year of studies he conducted a seminar. We were about ten students, and Aumann took us through the work of Schmeidler and Kohlberg on the nucleolus, which was an important breakthrough in those years. Every week Aumann would give us another set of problems, and those were tough problems. The following week, those who had managed to solve the problems would present their solutions. I remember that there were three of us who were competing neck and neck. We really had to work very hard and get into deep issues; the seminar was great! That's how I decided to do my M.Sc. with him, and later on, my Ph.D.

Q: What examples from your work best illustrate the use of game theory for foundational studies and/or applications?

A: I'm originally a mathematician, so most of my work is foundational. My work hasn't revolved around applications, even though they are extremely important. One cannot do game theory in a completely abstract way, without any roots in economics, biology, computer science, political science, and so on. If one wants to do something interesting, it is important for that something to be *relevant*! But I'm mainly on the foundational or theoretical side of game theory. My Ph.D.-thesis and some of my work throughout the years has been in cooperative game theory: axiomatizations, values, coalition formation. But I have also worked in noncooperative game theory, for instance, on repeated games. Since the nineties I have done much work with Andreu Mas-Colell on dynamic models. This includes adaptive heuristics and various dynamics, like regret matching, that lead to correlated equilibria and Nash equilibria.

Dynamic models are now quite an exciting area in which many people are working, so let me explain the general setup. Early game theory concepts were static concepts; Nash equilibrium, for example. Though there certainly is a dynamic intuition in the background, the definition is static, in the sense that this is a rest point. But that always leaves one asking how those equilibria are reached. If the players start at an equilibrium point, they will most probably stay there. But if they don't start at an equilibrium point, and they are reacting (say, best replying or better replying), or if it is an evolutionary process, then where does it all lead? Will it converge to this or that equilibrium, or not? These are difficult questions. While static analysis is relatively simple—not that it is simple, but it is *simpler than* dynamic analysis—the mathematics of dynamic systems is very complex.

On top of this, the class of interesting dynamics is huge. First, there are highly rational dynamics, where people observe what happens, calculate (say, the posterior probabilities), and optimize accordingly. This requires a lot of rationality and a lot of computation.

Second, there are dynamics that are essentially mechanistic and automatic, like evolutionary dynamics. In such setups there is no conscious computation or optimization by the participants. Instead, some process like natural selection makes the frequency of successful strategies increase. In addition, there are mutations – a kind of small random "noise" – that introduce new strategies. What happens is that those mutations that are successful keep expanding in the population.

And third, there are dynamics, like adaptive heuristics, where there is just a little rationality, that is, highly bounded rationality. The players act in simple and myopic ways that seem to be going in a "good" direction. Though the players are far from fully optimizing, in the long run their behavior may nevertheless yield the same outcomes that fully rational players would have achieved; take, for example, the simple adaptive heuristics of Hart–Mas-Colell that lead to correlated equilibria. So, to get back to your question, some of my recent work is on dynamics. But I have worked also on various other topics. For instance, I just gave a talk at the conference here in Stony Brook on the sure-thing principle and agreement theorems.

Q: For those who missed your talk, would you elaborate a little on this work?

A: It starts with joint work with Bob Aumann and Motty Perry. In decision theory, the famous sure-thing principle of Savage says the following. If I decide something when I know A, and I decide the same thing when I know B, then I should decide the same also when it is either A or B but I do not know which one. Now, that sounds exactly like the standard sure-thing principle of logic, right? In logic: if "A implies C" and "B implies C," then " 'A or B' implies C". *But it is not* the same thing, because in logic it does not matter whether A and B are compatible events or not, whereas in decision theory it turns out that it is essential that these events are *not compatible*, that is, that they are *disjoint*. If they are not disjoint, then the sure-thing principle of decision theory need not apply, and one may get into trouble using it. This is something that people don't realize, and it took the three of us quite some time and many arguments to put our finger on what exactly matters for the sure-thing principle, and why. And that came as a surprise. In fact, I gave a simple example to a full room here at my talk. At first everybody agreed that it was right, but then they saw that it was completely wrong. It takes people by surprise.

Aumann and I then went on to try to understand how to extend to decisions his famous agreement theorem that people with a common prior probability cannot disagree on their posterior probabilities when these are commonly known. Obtaining a decision-theoretic version of this result turns out to be conceptually difficult, though mathematically easy. One needs to understand what conditions are needed for it to apply. This, like much of my other work, is very foundational.

Q: Are there other examples of your work that you would like to mention?

A: Well, there is my work with Andreu Mas-Colell on the connection between strategic approaches and coalitional approaches. Let me explain. Game theory has two main branches: noncooperative and cooperative. The essential distinction is that noncooperative game theory deals with the strategic approach and the resulting strategic equilibria, whereas cooperative game theory asks questions such as what agreements players should reach. For instance, assume that there is a pile of money to split, and the players have the possibility of signing binding agreements—how will they split the money? To determine that, they will of course take into account what their other options are, in various coalitions. That's the focus of cooperative game theory: it is about coalitions and what outcomes the players should agree upon. Now an important issue here is how to make the connection between the noncooperative and cooperative branches. This is the classical "Nash program" started by John Nash; John Harsanyi is also instrumental here. It is about providing strategic foundations to cooperative game theory. Coalitions form, operate, decide, and divide the proceeds through strategic interactions between the players. The problem is that the strategic and bargaining procedures are not well defined; there are many possibilities to take into account, such as who proposes first and who second, what the rules for acceptance are, and how everything is conducted. Andreu Mas-Colell and I are trying to find procedures that are very general, yet give nice insights into what is happening. That's the direction our work took in the nineties, and we have just revived our interest in it (Andreu gave a talk about this here). It is an important direction, and I think that we have some new nice insights.

Q: What do you consider the proper role of game theory in relation to other disciplines?

A: Game theory is universal. It is relevant to all the disciplines in which people make decisions. It is also relevant to other disciplines, like evolutionary biology. Genes don't make decisions, but when you model evolution formally—the model of genes interacting, and *interacting* is the important issue here—then you see that it is a game-theoretic model. The insights and understandings of game theory become important here. But it also works the other way around: we game theorists learn from biologists. They develop something and we say, hey, that's a good idea. In fact, it has been a very fruitful connection in both directions. For instance, when the biologist John Maynard Smith introduced game theory into evolutionary biology in the seventies, it was quite rudimentary, but it caught on and flourished; then ideas from biology, like replicator dynamics and evolutionary approaches in general, came back to us and developed into what is now a very big area of game theory. Today the insights, tools, and concepts of evolutionary biology are used in game theory, and in economics too. Open *Econometrica* and you'll see not only theoretical papers that use evolutionary models, but, more generally, that the evolutionary paradigm is important and useful in economics as well.

Computer science is another relevant discipline. Computers make decisions too, and networks of computers interact, coordinate, or fight over resources; in short, game-theoretic problems need to be solved. The question of how to design the rules, the "protocols," in order to obtain desirable outcomes belongs to the game-theoretic area called "mechanism design." Nowadays, with all the electronic commerce, computer science is getting very heavily into game theory.

And one could go on. There is political science. There is philosophy. We have already touched upon logic and interactive epistemology. Basically, there are many, many areas to which game theory is relevant. It's like mathematics in the sense that a physicist uses mathematics to formulate what he is trying to model and explain. In the same way, an economist, an evolutionary biologist, or a political scientist uses game theory to formalize his insights and ideas. Game theory provides the tool for analyzing interactive situations, in which, unlike in simple optimization problems, one person alone cannot determine the outcome. What makes this a "game" is that what I do influences what happens to you, and what you do influences what happens to me. I thus have to take into account your rationality and your optimizing, and you have to take into account my rationality and my optimizing, and so on. At this point it all seems like one big mess, but game theory succeeds in cutting this Gordian knot; for example, by pointing to this or that kind of equilibrium. Game theory is thus a methodology that is applicable to the social sciences in general, as well as to other sciences like biology and computer science. I don't know if it is used in physics (but I recall some physicists in Jerusalem who presented a paper on a game-theoretic question). It is probably not used in any real sense in chemistry; I don't think that atoms are playing games—but one never knows. But in the social sciences, law, philosophy, biology, computer science, clearly game theory is an important tool.

I've compared game theory to mathematics, but it needs to be emphasized that game theory is not just a branch of mathematics. It is an applicable science, and as such relevant ideas and insights are essential. We had here at Stony Brook a "Nobel Session" two days ago. Bob Aumann and Tom Schelling, the Nobel Prize winners of last year, were both there. I gave an introduction to Aumann's work, and then Dick Zeckhauser gave an introduction to Schelling's work. At some point Zeckhauser quoted an article of Avinash Dixit on Schelling in the *Scandinavian Journal of Economics*, where he gives advice to the "budding economic theorist": do not just try "relaxing the condition of semi-strict quasi-concavity to hemi-demi-proper pseudo-concavity," but rather "obtain your primary motivation from life." I buy this statement completely. Taking existence theorems and merely improving the conditions here and there is not enough. (Of course, that doesn't mean that that should not be done. It is important when it allows us to expand the range of models and attack new problems.) You are not going to make a successful career out of doing *just that*. Ideas and concepts are essential. Intuition and understanding are essential. That was the point of Dixit and Zeckhauser.

But I want to emphasize that this is not the end of the story. It is *also essential to formalize* these ideas. Only by doing so do you realize that although your insights may sound convincing and look good, they do not give the full picture. If you just think of it conceptually you may miss many of those things. It is only when you try to prove formally what you think is the result of your model that you realize that it doesn't work; it might not work, for instance, because additional conditions are needed. Having beautiful and important insights and then being able to establish them formally is the right way to proceed. Game theory is not a "verbal" science. We are trying to be precise, because it's important to formalize the ideas. Aumann, for example, is a mathematician, but his great success does not come from just proving theorems. His success comes from taking a new concept, a new idea, and first of all being able to formalize it and strip it of all that is irrelevant, getting it to its barest. And then, when this is done, the results, theorems, and proofs usually become simple and clear. A significant part of Aumann's work, but not all, is like that; for example, the agreement theorem that we talked about. It's very important that you be able to formalize and to see whether your intuition is indeed correct. Now, if you have no intuition and you're just extending existing results, you're not headed for a successful career. It is the combination of intuition and formalization that is essential. That's an important lesson to learn. Now Schelling is not a mathematician and his contributions are mainly conceptual. But there is only one Schelling. And many people went on to formalize his beautiful insights by building the models and doing the math. As for Aumann, his forte is both on the conceptual side and on the mathematical side, and that's really the winning combination.

Q: That brings us back to the second question of good examples of game-theoretic work. That would mean that work like Aumann's not only uses mathematical tools but rests on new ideas and insights.

A: Yes, mathematics is a *tool*. You need an idea that is interesting and important and relevant, and you need to formalize it, which is the only way that you can verify that your intuition is correct. Verbal arguments are nice, but they are only the beginning; they cannot be the end. If you really want to verify that your idea is correct and that you're not missing something very important, then you have to be able to prove it formally.

Q: Would you touch on the famous quotation from the *Handbook of Game Theory*, where Aumann and you talk about game theory as a unified theory of the social sciences?

A: If I remember correctly, it says that game theory is like a unified theory for the *rational side* of the social sciences. It doesn't say "for the social sciences." For example, there is no claim that psychology is a subset of game theory. There is much in psychology that is not game theory, but at the same time there are areas of psychology, like rational decision-making, where game theory is of much relevance.

Q: But how far do these rational sides extend? That is, how far do the insights and ideas of game theory extend into the social sciences?

A: When you study rationality, you also study bounded rationality and irrationality. In addition to asking what happens in an ideal world where the decision maker is fully rational, you ask what happens when he is bounded—in his computations, in how much effort he makes, in his optimization. Beyond the ideal case of *homo rationalis*, there is a large part of game theory that deals with models of bounded rationality. It would be preposterous to say that everybody is fully rational. One should understand rationality and the limits of rationality. For example, what Danny Kahneman and Amos Tversky say is very relevant: it is important to understand under what circumstances we have biases and make errors in our decisions. Clearly we have biases; I know I do, and I try to correct them, but it's not easy. When I hear about a fifty percent discount, I tend to head for the store, but then I stop and think, wait a minute! I'm going to gain perhaps 10 dollars, or 50 dollars. Compare that amount to a one percent reduction in the price of the last house I bought. People will spend more effort on the former fifty percent than on the latter one percent, despite the fact that that one percent on the price of a house may be a huge amount, perhaps a few months' income. But they won't spend enough effort in trying to get that one percent reduction, by shopping around and bargaining. It doesn't come easily and naturally to us; we have to think about it. Also, we cannot always be the kind of rational person who computes everything. If you try to do that you'll be run over by the first car on the street. It's not a good survival strategy!

Q: What are the limits of rationality in behavioral economics?

A: People are definitely making mistakes when making decisions. They have biases and they make lots of errors. However, when it counts, when it *really* matters, we make far fewer mistakes. John List gave a beautiful example in an *Econometrica* paper a couple of years ago. He conducted his experiments on the floor of a sportscard market, where he found clear evidence of the so-called "endowment effect." This refers to valuing an object more when you own it than when you do not own it. Specifically, when people are given the choice of getting, say, a mug or a chocolate bar, about half the people take the cup and half the people take the bar. But if you give them, say, the mug, and ask them if they want to exchange it for the chocolate bar, very few agree; and the same if you give them the chocolate bar. Once they own it, they do not want to trade it for something else worth about the same. They value what they have more than the thing they don't have, while they are indifferent between the two objects when they do not own either of them. One needs to be careful in interpreting the results here, because they are all in a region where people are more or less indifferent. You might offer me a cup which is worth a few dollars or you might offer me a chocolate bar which is worth the same. I'll take the cup or the chocolate bar, I don't really care which; I just have to make a decision. And then keeping what I have when I am indifferent is a good and simple rule: "taking the path of least resistance." Also, in general I have more information on what I own than on what I don't own, and so risk aversion can explain this effect.

As I said, List found a very strong endowment effect among the people coming to the sportscard market. But then he also ran his experiment with the professional traders in that market—and, as it turned out, they had no endowment effect! The endowment effect vanishes perhaps because those traders who have it lose money and drop out of the market, or perhaps because they learn the hard way that that's not the way to trade. If you want to make money you had better make sure that you don't have biases that can hurt you ...

Q: ... when the stakes are high?

A: When the stakes are *real*, when they matter. We make errors all the time, we have many biases. But when it really matters, when you depend on it, you make far fewer errors.

Q: What do you consider the most neglected topics in late twentieth-century game theory?

A: Game theory is a dynamic subject. It evolves, and so it is hard to say that there are "neglected topics." There are so many people working in so many directions and new things are opening up all the time. There is no neglected topic in the sense that people should work on it and no one does. It's like evolution: if a topic is interesting, someone will work on it. Well, I will if no one else does. Of course, some topics are extremely difficult, we lack the tools to analyze others, and we don't know how to approach still others. It may take a lot of time until one starts developing the machinery.

For example, Harsanyi had a beautiful idea on how to deal

with games of incomplete information. Everybody wanted to analyze this clearly important topic, but nobody knew how. Then Harsanyi came up with the idea of transforming a game of incomplete information into a game of imperfect information by using "types." This allowed, inter alia, the development of a whole new field, that of (repeated) games of incomplete information, which subsequently bloomed. It's not that this field was neglected before; it was just too tough. It takes time until somebody finds an opening.

I'm not sure that I can point to a neglected topic. The boundary of the field keeps being pushed out. Now sometimes there are biases in science. Sometimes a topic that lots of people are working on may turn out not to be that important. For example, too much effort was put into equilibrium refinements, and they became quite esoteric at some point. It hasn't died out, but it has reached a certain maturity and is no longer as active as before. Evolution works beautifully in science. There are always many ideas, and the good ones spread and develop; "mutations" produce new ideas, and the good ones catch on, the bad ones die out.

Q: What are the most important problems in game theory today?

A: Dynamics is a crucial topic to understand. Most of our interactions are not static, one-time interactions, but rather repeated interactions. We interact day after day, whether with the same people or with different people. There may be repeated interactions, or things may change from one day to the next: the game, the people I play with, the environment, etc. All in all, dynamics is an important topic. A lot of work is being done now on dynamic models, and it is advancing very nicely.

Another area that is picking up quite significantly is the interface of game theory and computer science. With the Internet and all the electronic commerce, people realize that they have to understand game-theoretic notions, like mechanism design and auctions. Then there is also "algorithmic game theory," which deals with the problems of computing and finding equilibria, particularly in large games, like those involving big networks.

Speaking of applications, the latest *MIT Technology Review* selected the ten most significant emerging technologies of the year; "cognitive radio" was one of them. Think of all the cell phones, pagers, laptops, and wireless devices that everyone carries around nowadays. How can they use the bandwidth efficiently? Centralized allocation protocols are not practical; it has to be decentralized. Computer scientists have started looking into cognitive radio, programming each device to use a set of behavior rules that are based on game-theoretic ideas. Interestingly, the day after I found out about cognitive radio, I got an e-mail from another group of researchers working on something similar for cellular communication, who were trying to apply regret-matching algorithms from my work with Andreu Mas-Colell (but don't blame me if your cell phone stops working ...).

A further area of application has to do with congestion games, which arise in transportation problems. More and more cars are equipped nowadays with GPS navigation devices. Today these are mostly one-way machines: they get information on your location from satellites, and then compute your route. But there is no reason that it shouldn't be a two-way machine, where your GPS transmits your location. When a route becomes clogged with too many cars, the individual GPS devices could start routing cars to a different road. There are clearly problems due to the huge number of nodes—cars, intersections, routes—and the fact that they have to be solved online in real time. In all these areas game-theoretic dynamic approaches are very relevant. You could use adaptive dynamics that will lead the system to an equilibrium, or perhaps close to one. These are just a few of the many areas where game theory is very applicable.

# 10
# Ehud Kalai

James J. O'Connor Distinguished Professor of
Decision and Game Sciences
Kellogg School of Management
Professor of Mathematics
College of Arts and Sciences
Northwestern University, USA

---

Why were you initially drawn to game theory?

The 1960s were turbulent years on American campuses, as students were rebelling against the government, social institutions and "establishment" values. There was a push to increase awareness by the use of new approaches that go beyond the standard "linear thinking." As a PhD student in the mathematics department at Cornell University, I was affected by the spirit of the time, and by the feeling that work in pure mathematics could be isolating. So I decided to look for a career that would be more "relevant" and "interactive." My advisor, Jack Kiefer, suggested that I take a few courses that were more applied than the standard real analysis, algebra, and topology courses taught in the pure-math program.

Looking through Cornell's catalogue of courses, I marked some statistics and optimization courses but then noticed a course called "Game Theory," taught by William Lucas. The name of the course was puzzling and I decided to sit in on the course and see what it was about. It became "love at first sight."

I always loved the axiomatic approach and the logic of geometry, and here was a whole new application of this approach. Bill Lucas started with cooperative game theory, where simple appealing axioms turn out to be powerful enough to offer solutions to questions about individual behavior, group choices, fairness, and the like. It was surprising and fascinating. Although I cannot remember my feelings when I was introduced to geometry, I was amazed by the fact that you can resolve questions about fairness and behavior by simple mathematical axioms.

In addition to the fascinating process of "mathematizing" behavior, it quickly became clear that game theory was a young subject, and that many new ideas had not been tried yet. So creativity was welcome, with mathematics as its canvas. In addition to the wonderful teaching of Bill, young Louis Billera was around and was teaching many of us how to go about doing mathematical research in this subject. The enthusiasm of Bill and Lou, together with that of some of their visitors (including Guillermo Owen and Lloyd Shapley), was captivating, and indeed, during this period they produced some outstanding game theorists including Robert Bixby, Pradeep Dubey, and Robert Weber.

Game theory was still a small and unpopular area of research at American universities when I completed my PhD in 1972. But it was an active area of research in my home country, Israel. I took a faculty position in the Statistics Department of Tel Aviv University, where I could work with David Schmeidler (whose papers were central to my thesis), and arranged a visit with the leader of the Israeli school of game theory, Robert Aumann. The school of Israeli game theorists proved to be one of the most creative research groups. Joining it was a highly rewarding experience, both professionally and personally.

Another important step in my evolution as a game theorist was the move to the business school of Northwestern University in 1975. Game theory was still far from mainstream economics, but the young economists here, with strong support from the administration, were eager to incorporate it into economics. We became a major player in the "game theory revolution" in economics and management of the 1970s and '80s, and we are currently involved in similar revolutions in political science and operations management.

## What example(s) from your work (or the work of others) illustrates the use of game theory for foundational studies and/or applications?

At this stage of its development, it is useful to distinguish between theoretical applications and real-life applications of game theory. By "theoretical applications" I mean the contributions that game theory makes to other theories. Real-life applications are ones that address specific concrete problems faced by real decision makers.

Most of the applications of game theory until now have been theoretical. I refer to them as applications, despite the fact that they deal with theories, because what is a theory in one field may be an application in another. For example, economic theorists interested in the potential effect of a proposed government regulation study the Nash equilibrium of the game in which the regulation is imposed, and compare it with that of the game without the imposed regulation. Similarly, political scientists interested in comparing parliamentary systems study the games that result under the various parliamentary systems, and compare the efficiency of their game-theoretic solutions.

Game theory has been quite successful in dealing with theoretical applications like the ones above. There is almost no area of economics that was not drastically affected by game-theoretical tools. Similar revolutions are taking place in political and computer sciences. But even in areas not focused on human behavior, such as evolutionary biology, game theory has had a profound effect. At a more foundational level are the contributions of game theory to mathematics and philosophy. Interactive epistemology and the foundation of rational behavior provide a wonderful example. How do we mathematically model an individual's knowledge, including knowledge about the knowledge of others (which includes their knowledge of others, etc.)? What is rational behavior in situations that involve such complex knowledge structure? How does rational behavior respond to demonstrated violations of rational behavior?

I have made my share of contributions to game theory and to its applications to economics, social choice, operations research and computer science. But given that, unlike most game theorists, I have also had the experience of dealing with real-life applications; it may be useful to focus my discussion here on some of these.

As a business school professor, I teach game theory to business executives. Some of them come back to me later for consulting work, when they encounter strategic problems that require sophisticated analytical thinking. I will describe three examples that illustrate some of the benefits and the difficulties of two different game-theoretic methods, and their mixed record of success.

Two cases where game theory was applied successfully involved Baxter Healthcare, one of the world's largest health-care corporations. The first question involved the pricing strategy for Baxter's blood-plasma separating machines in China. The practice of Baxter and its competitor is to lend the machines, for free, to HMOs that use it, but to charge for the new plastic tubing needed for each individual blood donor. Chinese HMOs were gradually shifting to a strategy of taking the free machines from Baxter's competitor, but producing their own plastic tubing in China for substantially less than the price charged by the competitor. While Baxter itself was not cheated in this manner, market share was rapidly shifting to Baxter's competitor. Moreover, the competitor was not fighting this process of being cheated. It was not clear why the competitor lets the process continue, and how Baxter should respond.

Together with a team of experts (in finance, marketing, Chinese HMOs, and more) we developed a large strategy tree[1] that considered the major options available to Baxter and the other players (the competitor, Chinese HMOs, the Chinese government) over the next two years. The tree led to the conclusion that the best course of action was for Baxter to partner with local Chinese producers, and to produce their own tubing in China at a highly competitive price, even when compared with the local production by the Chinese HMOs. While the approach required making ad-hoc assumptions and simplifications of the problem, it had definite advantages. In addition to leading to a final solution to the concrete problem, it turned out to be a tool that facilitates communication within a large group of experts with diversified expertise.

Different experts contributed their input to their respective parts of the tree. For example, what Chinese HMOs might do in response to various actions of Baxter was assessed by the Chinese HMO expert. How the competitor might respond to various offers from Baxter (given that collusion is legal in China) was assessed by people who have negotiated with this competitor. The strategy tree was a vehicle for putting together all the information obtained from the various experts, and for discussing the various assumptions with management.

About one year later, Baxter's involvement in the production of distilled water gave rise to another game-theoretic study. Baxter's market share was being threatened by new entrants with new technologies. Even though the new technologies were at a very early

---

[1] I use the term "strategy tree" to denote an item between a decision tree and an extensive game. It is less sophisticated than an extensive game, because it does not include information sets. It is more sophisticated than a decision tree, since the major choices of opponents are explicitly modeled with some attention paid to the optimality of their choices.

stage, Baxter's strategy experts were concerned that not acting early could lead to a substantial loss in this highly profitable business. Some of the questions raised were the following: Should we buy off the competition? Should we partner with them? Should we compete with them and try to overtake them in the development and patenting of the new methods? If we compete, should we develop the technology and try to pass FDA regulation first in the US, first in Europe, first in Asia, or simultaneously in several locations?

This was more complicated than the China problem, since it involved a strategy of learning. (Does the new technology work? Can we assess, or even affect, the likelihood of FDA approval in the US by having it tested first in less demanding regions? Can we learn about the demand in the US by observing demands in other parts of the world?) Again, the method of constructing a large strategy tree proved effective. This time, our experts included engineers, FDA experts, marketing experts and more. At the end of several months of work, the tree and its assumptions were presented to upper management. The presentation was effective, and upper management decided not to ignore the new technologies and to devote several million dollars to study the issues in detail.

In addition to being effective, working with strategy trees is natural. In both of the above projects we interacted with two highly capable MBAs from Baxter. They caught on to the approach well, and our services were needed less and less as the work progressed. Unlike traditional game theory, the method involved much subjective optimization with a tree that was trimmed to the most essential issues at hand. We developed effective methods of sensitivity analysis that help in the construction and trimming of strategy trees.

The consulting work I describe next taught me lessons about the need for different methodology, but also pointed to additional difficulties in game-theoretic consulting. Arthur Andersen was a large accounting and consulting partnership headquartered in Chicago. It had two divisions, with hundreds of offices and thousands of partners around the world. Being a partnership, it used a formula to allocate its substantial profits to its partners. The profit-sharing formula was supposed to satisfy several objectives. It should be fair, so the partners would agree to it. It should create incentives to increase Arthur Andersen's profits, and it should have insurance aspects that prevent the income of partners from fluctuating too much due to the random annual fluctuations of the business in various geographical areas. I was hired by one of the senior partners to advise about the profit-sharing formula.

It became clear that the cooperative, rather than the strategic, game-theoretic approach would be more appropriate for the construction of a profit-sharing formula.[2] The monotonicity property of the Shapley value, which I studied earlier by myself and with Dov Samet, creates incentives for partners to exert efforts on behalf of the company. But it also became clear that computing the precise Shapley value of the problem at hand is (still) hopeless, and even if it were possible, the Shapley value theory is (still) not sufficiently developed. First, we needed a Shapley value applicable to a cooperative game with incomplete information. Second, the Arthur Andersen game was one that kept changing and evolving with time, so we needed a dynamic Shapley value with incomplete information. But these were not the only difficulties.

With a substantial use of *ad hoc* methods, I developed a Shapley-value-like formula for their dynamic game with incomplete information. However, the senior partner I worked for repeatedly put off considering my solution. It became clear that he really did not want a new solution, but only wanted to use me in order to justify the solution that they were using at the time. He wanted to shut down arguments from the opposing division, by stating that a game-theory expert specializing in bargaining and fairness approved his method. Despite the nice consulting fee I was collecting, I quit this job after learning an important lesson: Game theorists who think they have been hired to solve the game may turn out to be mere pawns in the game.

This phenomenon is not unique to Arthur Andersen. In many situations where companies hire consultants, it is often done for political and strategic considerations, rather than for the advice the consultant may offer. In game theory, where the consultant's recommendation typically affects players with competing objectives, the problem is more severe. We need to find ways to make the decision to hire the game theorist be independent of his recommendations.

---

[2] A strategic model would require a precise description of the moves available to all the players, the order of moves, the information available at the time of choosing moves, etc. This is clearly impossible.

## What is the proper role of game theory in relation to other disciplines?

In my roles as past president of the Game Theory Society and an editor of a game theory journal, I recently constructed a list of areas of activities in game theory and its applications. The list should give the reader an appreciation of the interactive nature of game theory.

1. **Non-cooperative game theory** studies the behavior of payoff-maximizing players who take into consideration all strategic and informational parameters.

2. **Cooperative game theory** studies how considerations of efficiency, fairness and stability guide the allocations of profits and costs to coalitions of rational players.

3. **Behavioral game theory** studies how real players play games: **experimental games** played in the lab, and **empirical games** played in the real world.

4. **Evolutionary game theory** studies play guided by imitation, survival of the fittest, etc.

5. **Algorithmic and artificial game theory** study issues of computational, informational, and behavioral complexity in games played by live players or by computing machines.

6. **Interactive epistemology** studies the subject of knowledge, including knowledge about knowledge.

7. **Combinatorial games** deal with mathematical issues unique to games.

8. **Non-Bayesian decision theory** concentrates on decision making under uncertainty, when relaxing or replacing the Bayesian assumptions made in the classical theory.

9. **Neurological studies of games** deal with physiological activities observed during the play of a game.

10. **Economic games** use the above tools to gain insights into strategic economic interaction and the performance of economic systems.

11. **Political games** use the above tools to gain insights into strategic political behavior and the performance of political and social systems.

Game theory has been described by some as the physics of the social sciences. I think that another useful analogy involves probability and statistics. Probability theory offers a language and rules for dealing with uncertainty, and statistics offers tools for real-world applications. These theories are designed to deal with uncertainty, no matter where it arises. Similarly, game theory offers a language and rules to deal with strategic interaction, wherever it arises. The Arthur Andersen and Baxter applications described earlier suggest an interesting parallel with statistics. When dealing with a real-life application, a statistician must choose the best among an available set of models: a classical approach, a nonparametric approach, a Bayesian approach, etc. An applied game theorist must choose between a coalitional model, a strategic model, or a hybrid of several models.

But thinking of game theory as similar to probability and statistics makes the relationship to other sciences clear. First, like probability theory, a well-developed game theory is foundational in any subject that deals with interaction. Second, in practical applications that involve strategic interaction, there is no way to avoid using game theory. However, following the practice of statistics, it may be necessary to have several different game-theory models, with the user choosing the appropriate model for the application.

As we have discussed, the interaction of game theory with economics over the previous century has concentrated mostly on theoretical applications. And indeed, in a similar manner to the use of probability theory, the use of game theory has become unavoidable in essentially all rigorous studies of strategic economic phenomena. We see that political science is going through a similar progression, even at a more fundamental level. Unlike economics, were much formal modeling was done prior to the arrival of game theory (through equilibrium models of supply and demand, for example), the initial formal modeling of political systems had to start with the use of game theory.

The interaction with evolutionary biology and computer science is interesting because there is a reciprocal fertilization between game theory and these other subjects. Since evolutionary biology studies interaction among species, it is natural to apply game theory there. But the reverse is also true. Evolution theories describe how the behavior of species evolves, without resorting to rationality but using concepts such as imitation, survival of the fittest, etc. An important finding there is that despite no reliance on rationality, species' behavior converges to what is predicted by Nash equilibrium. This is an important connection. It makes both theories more robust, and it is the subject of much current research.

Similar reciprocal fertilization is present in the interaction of game theory and computer science. Both game theory and computer science share a common goal: the mathematization of rational choices and behavior. Historically, however, computer science concentrated mostly on algorithms that generate rational choices, subject to complexity constraints. For a long time it ignored strategic and interactive aspects in systems that involved more than a single decision maker. Game theory, on the other hand, ignored issues of computation and complexity and concentrated mostly on the strategic interactive aspects. In recent years, we have seen a growing interaction and cross-fertilization between the two fields.

## What do you consider the most neglected topics and/or contributions in late 20th century game theory?

## What are the most important open problems in game theory and what are the prospects for progress?

What questions should a theory address? Questions should be sufficiently simple so that answers can be obtained, yet they should be sufficiently advanced so that the answers are meaningful. An important issue is the level of detail that is assumed as input to a theoretical model. In the case of game theory, where rational players are part of the model, the details must include the level of complexity that can be handled by the players. Game theorists refer to a model that assumes limited information or limited ability to handle complexity as one with bounded rationality.

A beautiful example of a theory of bounded rationality is the older subject of probability. Imagine going to a scientist before probability theory was invented, to ask for his advice on how to bet (H or T) on a coin. The fully rational answer is as follows. Find out the initial position of the coin, the rotational velocity, and the height to which the coin will be flipped, and the hardness of the material on which it will land. Then apply this data to the differential equations from physics that describe the trajectory of the coin. If you know all of these parameters and you are able to do the computation, you should know what side the coin will land on, and you should bet accordingly.

Clearly, the differential equations of physics described above are not the right tool to deal with the issue of betting on a coin. Both difficulties prevail: it is impossible to get all the necessary information, and it is impossible to do the computation, even if the information were available. Viewed in this light, probability is a wonderful theory of bounded rationality. Rather than attempting to answer the question of what side the coin will show, it addresses the question of what are the frequencies of H and T, if the coin is flipped many times. This latter question can be answered by experimentation, which leads to a reasonable recommendation for our decision maker.

In many respects, game theory is like decision theory before the introduction of probability. We are still wrestling with the identification of questions to answers, and how to go about the rational modeling of bounded rationality. For many applications, our models require the knowledge of too many parameters and assume unrealistic computational ability. For these reasons, we are often restricted to what social scientists refer to as stylized, or toy, models.

For older successful sciences, the identification of good models was guided by real-life applications. I think that game theory should move more in this direction. There are two related guiding questions: How to construct a game and how to play a game. The "how to play" question should go beyond 2 by 2 Prisoners' Dilemma games, and deal with more complex and realistic problems. (I do not mean to say that understanding the Prisoners' Dilemma game is unimportant—just as it was important for Newton to explain the effect of the law of gravity on an apple falling from a tree before moving to the movement of the stars. But the move from Prisoners' Dilemma to real life games may require substantial new approaches.) To outsiders who are not familiar with the communities of the researchers, it seems strange that the chess-playing program that defeated the best human player is of no interest to game theorists.

The chess example points to an area that needs addressing in game theory, which may be called "macro game theory." In chess, one uses *ad hoc* functions to rank board positions. For example, a position where a player controls more squares on the board is superior to one in which he controls fewer. It is better to have more pieces than the opponent. While not true in every board position, global considerations like these still guide the decision of chess players and programs.

Global considerations, like the ones in chess, are also used in our everyday decisions. For example, you are better off majoring in a subject that leads to more career opportunities. You are better off choosing actions that make friends, rather than enemies. While considerations like these are a part of our everyday real-life interactions, they are practically absent in strategic game theory.

Another important step for future game theory may be the development of additional sub areas and specializations, in a similar way to the development of statistics as a different subject from probability. While highly related, these subjects require different knowledge, tools and skills, and separation and specialization proves to be useful. A useful analogy is obtained by thinking of the construction of and flying of airplanes. It takes physicists to develop the underlying basic theory, aeronautic engineers to design a plane, and pilots to fly it.

While the engineers have basic knowledge of the physics involved, they have additional practical knowledge obtained by experimenting in wind tunnels and by learning about earlier plane designs. The pilots have basic knowledge of the physics and engineering, but have additionally needed skills.

The design and play of auctions may require decomposition similar to the one above. Game theorists offer the underlying theory. In addition to basic game theory, auction designers should have knowledge of behavioral theories, obtained in the lab and by studying the play of earlier auctions. Successful bidders understand the general principles and the rules of the auction, but have additional abilities to assess specific environments and opponents.

Recent criticism of game theory is that knowledge of game theory does not make one a better game player. While this may or may not be true, the airplane analogy is useful. It seems unlikely that a physicist would be good at flying a plane (or that a pilot could publish papers in physics), yet it does not mean that physics is useless in the construction of planes. Purely on knowledge of game theory, one is not likely to do well bidding in an auction. But this knowledge is useful for both the auction designer and the bidders.

It seems that game theory is starting to evolve in this direction. The development of the behavioral areas may constitute the first step in creating more practical knowledge, similar to the one of an aeronautical engineer. But it is important to recognize the need for these specializations and encourage, rather than resist them.

My hope is that this next century will lead to the development of more useful game theory. The problems addressed will continue to come from the social and biological sciences, but also from concrete real-life (as opposed to theoretical) applications. The methodology will rely on tools, concepts and ideas from other areas such as computer science, evolutionary biology, psychology, and others.

# 11
# David M. Kreps

Theodore J. Kreps Professor of Economics
and Senior Associate Dean

Graduate School of Business

Stanford University, USA

---

With permission of the editors, I'm framing my contribution to this volume as follows. First, I will give short answers to the "autobiographical" questions #1 and #2. Then, in lieu of answers to the more philosophical questions, I attach the text of remarks I made on January 6, 2007, to a luncheon hosted by the American Economic Association at the ASSA Meetings, held to honor the 2005 Nobel Laureates in Economics, Thomas Schelling and Robert Aumann. I have nearly unbounded admiration for the work of these two pioneers, and I think one can hardly do better than to examine their work, when trying to describe what work in game theory ought to look like, or what have been the most important accomplishments of the subject in the latter portion of the $20^{th}$ century, or what directions the subject should take now. For readers interested in a more direct version of my views on the subject, I suggest the Clarendon Lectures I gave a while ago, published as *Game Theory and Economic Modelling* (Oxford University Press, 1990); my remarks concerning Professors Schelling and Aumann will, I hope, provide a synopsis of those views, while at the same time giving an account of the contributions of these two giants. I close by briefly connecting (stylistically) my own modest work to that of Schelling and Aumann.

## Why were you initially drawn to game theory?

As a graduate student at Stanford University and, subsequently, as an Assistant Professor in the business school there, I was privileged to be able to attend the summer meetings of the Institute for Mathematical Studies in the Social Sciences, the IMSSS. Among the perennial "principals" of this summer-long series of seminars were Ken Arrow, Mordechai Kurz, Frank Hahn, Bob Aumann, Eytan Sheshinski, and Robert Wilson; Michael Rothschild, Joe Stiglitz, and Mike Spence were there at least some of the time; Gerard Debreu, Roy Radner, and John Harsanyi were frequent visitors from Berkeley. It seemed that the whole world of economic theory passed through, and I got to sit in the back of the room and drink it all in.

To be sure, nothing that I was working on at the time could be called game theory. My interests (in terms of research) have always centered on topics connected to dynamic choice and dynamic behavior. As an undergraduate (before coming to Stanford), both in my senior thesis and at summer jobs at Bell Laboratories, I worked on optimal stopping problems; my Ph.D.-dissertation in Operations Research concerned the theory of dynamic programming. This led me to work in two directions, axiomatic choice theory – emphasizing dynamic choice – and dynamic models of financial markets. But, at the same time, I was fortunate to be able to see and hear a lot of game theory and, in particular, the first connections being made between noncooperative game theory and substantive issues raised in information economics. It was an exciting time, and when, a few years later, I began work on problems connected to game theory, it seemed a natural intellectual progression.

## What example(s) from your work (or the work of others) illustrates the use of game theory for foundational studies and/or applications?

I've employed the concepts and techniques of game theory in a variety of papers, on a variety of topics, but I suppose three applications stand out.

The first concerns work I did with Robert Wilson, simultaneously developed by Paul Milgrom and John Roberts, on models of reputation in repeated games with incomplete information. (We subsequently adapted the basic ideas into a four-author paper that used the prisoners' dilemma. The first paper is "Reputation and

Incomplete Information," *Journal of Economic Theory*, Vol. 27 (1982), 253–79. The second is "Rational Cooperation in the Finitely Repeated Prisoners' Dilemma," *JET*, Vol. 27 (1982), 245–52.) When Party A is uncertain about what motivates Party B, and if Party B has lots of opportunities to demonstrate to Party A what are her apparent motivations, (how) can Party B use this to her advantage? What if Parties A and B interact repeatedly and each is unsure about the motivations of the other? What if Party B interacts with a series of opponents? What if B interacts with a number of opponents through time, but simultaneously? (The last concerns work done later with Drew Fudenberg, in "Reputation in the Simultaneous Play of Multiple Opponents," *Rev. Econ. Studies*, Vol. 54 [1987], 541–68.) Our early papers showed that, under the right circumstances, a little uncertainty can go a long way in changing the nature of equilibrium interactions. (It was in the context of writing these papers that Wilson and I wrote "Sequential Equilibria," *Econometrica*, Vol. 50 (1982), 963–94, which is our most cited paper; its purpose was to give us the tools needed to study the contextual problem of reputation in the presence of incomplete information.)

The second ("Corporate Culture and Economic Theory," in Alt and Shepsle [eds.], *Perspectives on Positive Political Economy*, Cambridge University Press, 1990) involves the application of Folk Theorem ideas to the theory of organizations and, in particular, to the question: What is the role of a "firm?" Among the many and varied valid answers to this question, my work explores the idea that a firm's identity establishes, in essence, a long-run mode of equilibrium behavior amongst the "members" of the firm. Among the facets of this that are explored are two of particular importance:

- The contingencies faced by employees of the firm over the course of their time there cannot all be anticipated *ex ante*, and so the "mode of equilibrium behavior" has to take the form of a general culture or rule of thumb for how contingencies are met. To preserve the "equilibrium-ness" of this requires that, at least, participants are able to say with some reasonable measure of consensus whether the adaptations employed in meeting a particular set of contingencies are or are not in accordance with the culture or rule of thumb. This both places positive value on cultures that are relatively simple and straightforward and/or in accord with broader social norms and forms of behavior – they are less susceptible to being misunderstood or misevaluated *ex post* – and on an organizational focus that limits the range of contingencies that must be met.

- Insofar as "last moves" in particular encounters exist, the decision rights for making those last moves should belong to parties with an ongoing stake in living up to the organizational culture or rule of thumb. The stake is perhaps pre-eminently going to be a reputation stake, which may reside in the individual decision maker or, if the decision maker shares in the value of the ongoing enterprise, in the firm's collective reputation. We can imagine, say, a series of short-lived "bosses" who must, in the style of Herbert Simon's hierarchical superior in his seminal paper "The Nature of the Employment Relationship," make decisions that affect the welfare of short-lived employees, if two conditions are met: (a) the decisions of the bosses are relatively widely known to future prospective employees of the firm (i.e., if the firm has a clear culture, and decisions in violation of that culture are known to others), and (b) if the employer, although short-lived, retains a financial stake in the value of the enterprise after her decision is made, so that in making the decision, she has the incentive to preserve the enterprise's value by conforming to its cultural norms.

The third piece of "research" I would cite is, I suspect, largely unknown to economists. Indeed, most economists would probably not classify it as research at all, as it is not journal research but instead a MBA-level textbook, with nary an equation in it. Coauthored with James Baron, *Human Resource Management: A Framework for General Manager* (J. Wiley & Sons, 1999) is an attempt to integrate a game-theoretic/transaction cost economics approach to employment with insights from social psychology and organizational sociology. Building from Simon's primitive model of the employment relationship as a long-term relational contract in which the hierarchical superior retains most of the decision rights, we use social psychology and organizational sociology to help us "fill in" missing data about how employees form their expectations and evaluate their outcomes. For instance, we explore how social comparisons, attribution processes (specifically, processes of under- and over-sufficient justification for effort), and gift exchange can all turn some of the predictions of agency theory on their sides, if not their heads. (In case you are curious and wish to avoid a book, a paper-length synopsis of the book is being prepared as I write this, to be entitled "Employment as an Economic and a Social Relation," to appear someday in Gibbons and Roberts [eds.], *The Handbook of Organization Theory*, Princeton University Press.)

Readers familiar with these three applications may object that while the first is surely an example of applied game theory, the second and third, with concepts such as unforeseen contingencies and psychological processes of gift exchange floating around, are not. This is a point to which I will return at the end of this essay, after reproducing my remarks in celebration of Aumann and Schelling.

───────── ♦ ─────────

## Remarks in Celebration of Nobel Laureates Robert Aumann and Thomas Schelling

Delivered at the AEA Luncheon in Their Honor, January 6, 2007

Ladies and Gentlemen, Professor Schelling, Professor Sargent ...

It is both an honor and pleasure to have this opportunity to sing the praises of the two winners of the 2005 Bank of Sweden Prize in Economic Sciences in Memory of Alfred Nobel, namely Professors Robert J. Aumann and Thomas C. Schelling.

I begin with the official citation of the Royal Swedish Academy, which entitled the prize *Conflict and cooperation through the lens of game theory*. The citation reads:

> *Why do some groups of individuals, organizations, and countries succeed in promoting cooperation, while others suffer from conflict? The work of Robert Aumann and Thomas Schelling has established game theory – or interactive decision theory – as the dominant approach to this age-old question.*

Now, that's saying something. Nearly everything in economics is ultimately about cooperation and conflict, but more than that: Economics, Sociology, Political Science, Anthropology, and many parts of Psychology concern cooperation and conflict. So if indeed the work of Aumann and Schelling established game theory as the dominant approach to the study of cooperation and conflict, they surely earned the Nobel Prize.

Just so there is no misunderstanding, let me say at the outset: While I dislike formulations of this or that approach as *the* dominant approach to anything, I fully believe that these two individuals played a pivotal role in establishing game theory as a much-used ... and enormously useful ... tool in the study of these questions and many others, utterly changing economics as well as other social sciences. For their general role in this methodological revolution, as well as for the specific contributions they made, they bring honor to the prize.

But I am by nature a skeptic, and I think it is worthwhile to ask and answer two questions, before we grant the truth of this citation:

- What, specifically, did Aumann and Schelling do, to contribute to our understanding of the question? Why do some groups succeed in promoting cooperation, while others suffer from conflict?

- And, second, did they indeed establish game theory as a central approach to this question? How did they manage this?

The first question is, of course, the easier of the two. In fact, the Swedish Academy, in its *Advanced Information* issued with the announcement of the prize, tells us what they saw as the specific contributions of the two laureates: Professor Schelling is specifically cited for a variety of insights and concepts concerning conflict, commitment, and coordination, and in particular for the application of these insights and concepts to issues of deterrence and the arms race. These contributions are collected in Schelling's book *The Strategy of Conflict*, which I feel relatively confident in saying has been the most widely read and influential book about game theory in the literature. The book abounds with specific contributions, but to name three:

- In the context of aggression and deterrence, modeled with the game of chicken or hawk-dove, Schelling studies how the ability credibly to commit to respond in the event of aggression or, even more efficiently, to respond with sufficiently high probability, can lead to effective deterrence.

- Schelling discusses how, by taking a complex and seemingly intractable bargaining issue a step at a time, the parties may be able effectively and efficiently to resolve the issue.

- And, for situations where coordination is key, his notion of a focal point provides tools that are sometimes handy for resolving problems of a multiplicity of equilibria in a particular situation.

What Schelling does in *The Strategy of Conflict* is to analyze truly significant contexts, laying out for the reader the strategic nub of the situation. Perhaps I can illustrate the nature of his analyses by returning to the first sentence of the Nobel Citation:

> *Why do some groups of individuals, organizations, and countries succeed in promoting cooperation, while others suffer from conflict?*

This citation puts cooperation in opposition to conflict. But are they opposites? Schelling, examining this question, would show how, in some contexts, some forms of conflict, managed institutionally or within limits, can *promote* long-term cooperation. He wouldn't accept the simple view that they are opposites but, working through specific contexts, tease out a host of subtleties in how conflict relates to cooperation.

Professor Aumann is cited specifically for his pioneering work on the formalization of the Folk Theorem. I'm pretty sure that no one in this audience needs me to recount the Folk Theorem, but I'll do so anyway, to be complete. It says that, if a number of parties interact repeatedly, many patterns of behavior, including patterns in which parties sacrifice short-run self-interest for longer-term cooperation, can be self-enforcing arrangements or Nash equilibria. The idea is that each party is willing to sacrifice her own short-run interests, because she is promised good treatment by others if she behaves, or she is threatened with punishment by others if she does not.

This is not, of course, some brilliant and until-then unrecognized discovery. It is something that children are meant to learn in nursery school, where it is called learning to share and to take turns.

Aumann's specific contribution, according to the Swedish Academy, was to take this piece of folk wisdom and formalize it. Aumann, alone and in collaboration with others such as Michael Maschler, Richard Stearns, and Lloyd Shapley, and with further contributions made by Ariel Rubinstein, Drew Fudenberg, Eric Maskin, and others, made the result precise:

- They establish which outcomes can be equilibria and which

not, depending on long-run decision criteria of the participants.

- They establish the precise meaning of equilibrium here: Are these Nash equilibria, or strong? Are they perfect?

- And they explore how various structural parameters – the number of participants, observability of actions, frequency of interaction, the level of completeness of information, even the computational capabilities of the players – affect what can and cannot be achieved.

To illustrate what we gain from these formal investigations, I'll criticize the characterization I gave moments ago of the Folk Theorem. I spoke in terms of playground behavior that involves learning to share and take turns. It is natural to think that the Folk Theorem is about how long-term considerations allow parties to move from conflict to cooperation. When presenting it for the first time, who has not used as examples repeated play of the Prisoners' Dilemma or, in economics courses, oligopolists who gain the ability to collude?

But Aumann's formal work makes it clear: The Folk Theorem is about which payoffs are rendered strategically feasible by repetition, not which category of behavior. In other contexts, repeated play and the logic of the Folk Theorem can render conflict and aggression strategically feasible, where short-run considerations rule it out. We might like to think that playgrounds are where children learn to share and take turns, but it is also where they can learn how to bully. By being precise about what repeated play enables, Aumann opens our eyes to the full range of possibilities of this sort of thinking.

These are not, of course, the only contributions made by the two laureates, over their lifetimes of writing and research. The Nobel Prize is meant to be for specific contributions, and not as a lifetime achievement award. But it would be wrong not to mention, at least in passing, other contributions of the two laureates.

Schelling, in *Micromotives and Macrobehavior*, writes about how the seemingly insignificant actions of many individual agents can, in some circumstances, have profound social consequences, because they feed on one another. And, following earlier work by Strotz, in *Choice and Consequence* he explores issues of self-control and the challenges that lack of self-control may pose for policy makers.

Aumann has made a huge variety of contributions to the mathematical foundations of game theory and general equilibrium. This includes work on the value of a game; on epistemics and the epistemic foundations of equilibrium, perhaps most significantly concerning the notion of common knowledge and the related concept of correlated equilibrium; on economies with a continuum of agents; and on choice under uncertainty.

And their contributions extend beyond their own work. Both Schelling and Aumann have had profound influence on the profession through the many students they trained, and as exemplars for the diverse styles but uniformly high quality of their research.

Now let me comment on their styles of research. Looking first at the specific research cited by the Nobel Prize Foundation, but looking even more at their lifetime portfolios of contributions and at the work of their students, it seems apparent that these are two very different sorts of economists, making two very different sorts of contributions.

Put more starkly than is warranted, I find it hard to imagine a paper by Professor Aumann that doesn't contain at least one very general mathematical theorem, and I find it equally hard to imagine a paper by Professor Schelling that does.

I hasten to add: This, I think, is starker than is warranted by the facts; it mischaracterizes both individuals. I have had the pleasure – and it is a true pleasure – of hearing Professor Aumann give a number of papers. And, to the extent that my memory can still be trusted, his style is always the same. He begins with a simple story involving, at least in English, the problems faced by a pair of individuals, Alice and Bob. Alice and Bob's problems invariably begin as relatively simple, but they quickly get ... well, complicated isn't the right word ... they become subtle. Aumann, however, having begun with simple and intuitive problems, resolves the subtleties that he introduces with equal intuition. The subtleties lead him to a fairly abstract general idea and a very general mathematical theorem, but one understands the intuition behind the theorem because of the path taken deriving it.

I have not, on the other hand, ever had the pleasure of hearing Professor Schelling in person. But I have read his books, and he also works from specific example to more general concepts, starting with relatively simple problems that become increasingly subtle. He also develops the reader's intuition as he proceeds. He is less likely than Aumann to pull his threads together into a mathematical theorem, but he presents a general theory nonetheless, just one that, in most instances, is more verbal.

And, while Aumann likes to end his presentations and papers with everything neatly tied up, Schelling – at least in my reading – likes to end each of his chapters letting the reader know that the story doesn't end here. He concludes with a final twist to the story, keeping the reader a bit off balance and clear that the theory developed is, at best, an interim statement of what is what.

I found it interesting, in preparing these remarks, to read how others react to these differences in style. Much is made, at least in the commentary I read, about how different the two are, and how different are their contributions. Indeed, perhaps the most interesting comment concerning the differences came from Professor Schelling himself. He was quoted by the BBC and elsewhere, on the day of the prize announcement, as follows:

> ... they [the Nobel committee] linked us together because he [Aumann] is a producer of game theory and I am a user of game theory.

This is hardly the occasion to pick a fight, least of all with one of the individuals I am called upon to praise, but I am afraid that this is a case where I must disagree. Professor Schelling, I don't think you've quite captured the link between your work and that of Professor Aumann.

To explain how I see the link, I return to the second sentence of the Nobel Prize citation. Let me remind you. Rewriting a bit, it was that:

> The work of Robert Aumann and Thomas Schelling has established game theory – or interactive decision theory – as the dominant approach to the age-old question of why some groups promote cooperation while others fall prey to conflict.

Is this true? Is game theory now the dominant approach to this age-old question? Why is it, if so, and how did the work of the two laureates make it so?

First, as I said earlier, I am always uncomfortable with assertions about some approach being *the* dominant approach to hugely important questions. I think it is dangerous, given any individual's limited breadth of knowledge, to make that sort of claim. So I won't go that far.

Certainly, game theory has, over the past twenty to thirty years, been a very fashionable approach to this question, by economists, as well by at least some sub-communities of political science and sociology. Now the word "fashionable" may suggest to you a hidden criticism, and I mean none. I simply mean that this is an approach that has gained a lot of adherents among social scientists. But I'll try to defuse the notion that I'm being critical by amending this somewhat: I believe that game theory has been a very fashionable ... and entirely serviceable and immensely useful ... approach to this question.

Why has game theory been fashionable and, more importantly, serviceable and useful? There are, I think, three intertwined reasons:

1. Game theory, as a language or form for modeling multi-person interactions, has proved to be a fairly flexible and yet precise modeling language. As with any formal and, to some extent, mathematical modeling language, it permits the person doing the modeling to trace from assumptions about human behavior and institutions to conclusions about what will ensue. It allows others to comprehend what the modeler is assuming and how those assumptions do or do not lead to conclusions. It allows modelers to comprehend similarities and differences in various contexts – how the problems of a nation state trying to convince a rival that it will retaliate, of a central banker trying to convince the population that he will not inflate the currency, and of a individual trying to convince himself not to smoke a cigarette when it is late at night are or are not the same – so that insights in one context can be transferred to another.

2. It is particularly good at moving one's attention from static considerations to dynamics. If you go back to the specific contributions of both laureates, it is obvious that these are contributions that turn hugely on dynamic interactions. To say a word more here, game theory has given us good modeling tools for moving from the no-doubt important study of spot transactions to the broader study of economic, social, and political *relationships*, be they between individuals, organizations, or nation states.

3. Because of the importance of the issues attacked and the insights generated, its application to specific issues has attracted the attention and interest of a wide audience in not only the economics community, but throughout social science.

To be sure, game theory is a language or form for modeling. It succeeds or fails based on how well it is applied in specific contexts, and not because it provides answers to specific questions, *per se*. I recall that, at the time of the announcement of this prize, the press asked experts in the field to identify examples of where game theory had provided insights. One expert cited the Oslo Peace Accords. He said that game theory bore out Schelling's insights into how breaking a complex and seemingly intractable issue into pieces could lead to success. Simultaneously, another expert also cited the Oslo Peace Accords. Game theory, this expert said, showed how taking issues piecemeal could be a recipe for failure and disaster. Of course, game theory, *per se*, shows neither thing. Models built using the language of game theory show how, *depending on assumptions about behavior of the individuals involved*, either conclusion might be valid.

More generally, to the extent that we are focused here to some extent on applications that employ the Folk Theorem, one must understand that, in specific situations, predictions will not be anything like "the point where supply intersects demand." The entire point of Aumann's theorems is that, with repeated play, many things may be equilibria, and of course the same is true in the sorts of coordination games that Schelling studies with the concept of a focal point.

But is this really so different from "supply equals demand?" The point where supply equals demand is determined only after supply and demand are determined; the precise equilibrium depends on specific tastes and technologies. In these sorts of settings, we add beliefs, or expectations, to that list. But when one moves from static spot exchange to dynamic relationships, of course beliefs and expectations today will be important determinants of equilibria. It is a *strength* of game theory that it makes this dependence clear, telling us what more we must know to be able to predict outcomes.

In any event, the strengths of game theory as a form or language for modeling have made it both fashionable and extraordinarily useful, and Schelling and Aumann played a huge role in making this so. As the Nobel Prize Committee pointed out, for game theory to have reached this position, it took the right tools to be developed and for researchers to show how useful those tools could be. The first game-theory prize recognized the development of the right tools: John Nash's basic equilibrium concept; Reinhard Selten's concepts of perfection, which are crucial to the analysis of dynamic games; and John Harsanyi's games of incomplete information, which allow us to study situations of private information. This prize, to Aumann and Schelling, recognizes pioneers in the analysis of interesting economic and social situations.

Professor Schelling asserts that he is a user of game theory, and indeed he is. But he used game theory to study important issues in novel, insightful, and provocative ways. Because the issues he illuminated were so important and his insights so deep, he drew others into the conversation that he began. By doing this, he helped change the direction of game theory itself: by changing the tone and direction of the conversation by creating and using new vocabulary, he had a profound impact on the language. So while he is indeed a user of the theory, his uses shaped the theory, making him an important creator of the subject, as well.

Aumann has also had deep insights into specific questions. With his contributions to common knowledge and agreeing to disagree and in the idea of a correlated equilibrium, he certainly added powerfully to the vocabulary of game theory. But he is recognized here for something a bit different. By establishing rigorously what the Folk Theorem says and what it doesn't say, and in other important general results he proved, he established how powerful and general the new vocabulary forms could be and where they were limited, making those forms more accessible and, ultimately, useful to the wider community.

I don't know that I'm fully sold on metaphor, because virgin territory is rarely a reality, but I think the relationship between Schelling's work and Aumann's can be described in terms of settling hypothetical virgin territory: Schelling is an explorer of *terra incognita*, who explores the new world without the benefit of maps or guides, showing us new and exciting possibilities. Aumann, on the other hand, is someone who breaks the trails and composes the guides that the rest of us use in settling this new world. They both helped to shape the subject and, perhaps more importantly, each in his own way brought this way of thinking to the forefront of the disciplines of social science. This is no mean feat, and they really did it.

That's a great place to stop, but – perhaps trying to imitate Schelling – I'm going to close by saying that it is *not* the end of the story. Game theory has been a great success because it gives us the forms and language that allow us to discuss issues like reciprocity and commitment. But there are pieces of the contextual puzzles of cooperation and conflict and more generally of dynamic social relationships, with which game theory, so far, has coped less successfully. To continue to be serviceable and useful, game theory must continue to grow and develop.

I don't know if the two laureates, or anyone else, for that matter, will agree, but I think we need in particular to get better at modeling human behavior, especially in dynamic situations, and at understanding how individual beliefs and even tastes are formed and reformed through time.

Some of this concerns the territory already explored by Professor Schelling, but with which we have not really come to terms, such as a full understanding of focal points. Some concerns issues that have been explored by Professor Aumann but which are still unsettled, such as the problem of inference in the face of observations that are counter-theoretical. And there is still a lot of terra incognita which only now pioneers in behavioral economics are beginning to explore. Aumann and Schelling – and Harsanyi, Nash, and Selten – have given us a marvelous methodology for dealing with issues of cooperation and conflict and other sorts of social relationships. But the methodology is a work in process. To do justice to the enormous contributions of these two giants, and to honor them appropriately, we must stand on their shoulders and continue to explore and domesticate, in the best traditions of each.

───────────────── ♦ ─────────────────

## Postscript: Back to the three examples of my own work

With the truly seminal contributions of Schelling and Aumann establishing the gold standard for work in game theory, it may help to say a few words more about the three streams of my own work that I mentioned. If I flatter myself, the work on reputation and incomplete information, and the related development of the concept of a sequential equilibrium, is in the tradition of Aumann. This work is, for one thing, highly mathematical. I hope it is not merely mathematical – that it has a strong intuitive and applications-based core – but, in the end, the results are mathematical propositions (if not quite theorems). To be truly in the tradition of Aumann, Wilson and I would have had to have pushed the results further towards generality; in the end, we were only looking at some illustrative examples. Fudenberg and Levine, "Reputation and Equilibrium Selection in Games with a Single Long-Run

Player," *Econometrica*, Vol. 57 (1989), 759–78, does the heavy lifting of getting general results for at least some important cases.

The other two streams of work, on the role of the firm as a "bearer" of culture or implicit contracts/agreements and on human resource management, are closer to the style of Schelling. They explore concepts such as unforeseen contingencies, gift exchange, and social comparisons, which are not (yet) put into rigorous or general mathematical terms. I don't believe it is impossible or even that difficult to begin to treat these ideas more formally; in the case of unforeseen contingencies, I've even made what I hope is a start (in "Static Choice in the Presence of Unforeseen Contingencies," in P. Dasgupta, D. Gale, O. Hart, and E. Maskin, [eds.], *Economic Analysis of Markets and Games*, MIT Press, 1992). But it hasn't been done yet, and it is precisely this sort of work that I am referring to in the next-to-last paragraph of the remarks.

# 12
# Herve Moulin

George Peterkin Professor of Economic Theory

Rice University, USA

---

**Why were you initially drawn to game theory?**

In the early seventies I was a student of mathematics interested in its applications to the social sciences but not trained in economics. The easiest route toward these applications goes through Game Theory: it is built upon a transparent behavioral model, namely rational choice in its most abstract form, and is readily accessible to anyone with a taste for mathematical modeling.

Its seminal result is the beautiful minimax theorem, both a non trivial mathematical statement and one with extremely intuitive applications, from penalty kicks in soccer to the inspection policy at customs. More in Question 3 about this theorem.

The next level of modeling involves games where the interests of the players are only partially antagonistic, and there too the relevance to real life is immediately clear, as I discovered in Thomas Schelling's *Strategy of Conflict*. However it was obvious then as it is now that in a general n-person game, it will not be possible to advise the participants about an "optimal" strategy as long as we stick to a context-free interpretation of the game situation. That the analysis of $n$-person games is an open ended conceptual quest was already clear in the subtle discussions of Schelling's book. Further evidence was in the best Game Theory textbook at the time, by Guillermo Owen, discussing side by side the cooperative and non-cooperative models without relating them to each other in a logically satisfying way.

I jumped happily into that gigantic conceptual puzzle, with an eye on the interpretations of the game most relevant to economics and other social sciences.

## What examples illustrate the use of game theory for foundational studies and applications?

Forty years or so after Von Neumann and Morgenstern's intuition, Game Theory has systematically reshaped the entire field of economic theory by providing analytical tools that encompass classic price theory and Walrasian competition, and go much beyond. Probably the most dramatic impact is in the field of Industrial Organization, where economists were finally able to build upon the seminal work of Cournot and Bertrand in the $19^{th}$ century, and develop an integrated model of competition between firms in which monopoly and perfect competition are just the two limit cases.

In my own research I have been guided by three outstanding contributions of Game Theory that raise many fascinating theoretical questions and are motivated by countless applications in the social sciences.

- *core stability*

When the allocation of scarce commodities is governed by the regime of private property, the market participants are free to engage in any trade they please. Imagine that a benevolent arbitrator wishes to propose a socially optimal allocation (specifying the share of every agent) that will be agreeable to all parties. If a coalition of agents can do better by ignoring the proposal and trading their own resources, the allocation proposed by the arbitrator is not sustainable because of the *objection* in question. Those allocations against which no such objection is available to any coalition are said to be *core stable*, and their set is called the *core* of the economy. The core is to cooperative games what the Nash equilibrium is to non cooperative ones, namely the main concept of strategic stability.

A celebrated theorem due to Gerard Debreu and Herbert Scarf says that, when each agent owns a vanishingly small fraction of the resources in the exchange economy, the core coincides with the set of competitive (Walrasian) allocations. Thus the purely decentralized price-taking behavior captured by the competitive equilibrium and the radically different story of direct negotiation within coalitions of any size point to the same set of allocations.

Competitive Walrasian analysis can only be developed when we trade divisible commodities and production technologies have increasing marginal costs. Core analysis in general cooperative games ignores such limiting assumptions. It can even be used for allocation problems where efficient trading cannot be orchestrated by a price signal such as the bilateral matching problem, where each agent on one side of the market seeks the best match on the other side. The Gale-Shapley algorithm, which is successfully applied to match hospitals to medical residents and students to high schools, computes efficiently the core stable matching most favorable to one side of the market.

Identifying classes of resource allocation problems where core stable outcomes always exist is one of Game Theory's activities most immediately useful to economists.

- *the Shapley value*

In many fair division problems, simple ideas such as equal split of the benefits or proportional division according to some unidimensional parameter do not work. A typical example is the allocation of common costs involving complex externalities. Think of dividing the profits of a joint venture to which three partners bring different skills (one is a lawyer, another an engineer, the third is an IT specialist and a translator) and on which they work different numbers of hours, the lawyer from his regular office, the engineer by visiting clients, and the IT person online. To cut the Gordian knot of these intertwined externalities, the Shapley value offers a broadly applicable solution, based on *counterfactual* profits: how much profit could the lawyer and the engineer garner without the IT specialist (buying IT and translation services on the market)? How much profit for the lawyer and the IT person (hiring an engineer on the market)? And so on.

The Shapley value computes a certain linear combination of these imaginary profits and of the actual profit to be divided. Although other solutions can be designed from the same list of counterfactual profits, the Shapley value stands out for its relative computational simplicity and the strength of its normative justifications (see Question 4 below).

- *strategy-proofness*

Suppose a community must choose the amount of taxes it wishes to raise for building a swimming pool. In order to compute the socially optimal expense for the pool, the town manager elicits the preferences of the residents, namely their willingness to pay for pools of various sizes, then aggregates the answers in some fashion to reach a decision. A fundamental difficulty of this procedure is the incentive compatibility issue: if residents who declare that they do not care for swimming are not taxed, an avid swimmer will be tempted to *free ride* by reporting zero willingness to pay, and showing up anyway to swim once the pool is built and paid for. Sharing the tax burden equally among all residents takes care of this kind of free-riding but may create the opposite incentive: our swimmer may influence the choice of a bigger pool by exaggerating her willingness to pay, thus compensating for the low enthusiasm of other residents.

Designing allocation procedures in which participants have every reason to report truthfully their preferences and other personal characteristics that they are the only ones to know has been the central question of the mechanism design literature for over 30 years. The key concept is that of a strategy-proof mechanism, such that it is a dominant strategy for each participant to be truthful. Characterizing strategy-proof mechanisms in a given allocation problem is typically a hard mathematical question. In the few simple problems where the question can be solved, it leads to procedures immediately applicable. Examples include the median voter compromise when preferences are single-peaked; Vickrey's second price auction and its generalization as the family of Vickrey-Clarke-Groves mechanisms; Shapley and Scarf's algorithm for the assignment problem and serial cost sharing.

## What is the proper role of game theory in relation to other disciplines?

Game Theory started historically as a branch of applied mathematics. Today a small but significant fraction of the discipline is developed by scholars who could just as well be (and some times are) housed in departments of Operations Research, Statistics, Applied Mathematics, and more recently Computer Sciences.

The relation between Mathematics and Game Theory has been fruitful for some time. Several points of contact jump to mind immediately: fixed point theorems in functional analysis, mathematical logic and the theory of cardinals; dynamical systems, learning and evolutionary models; complexity issues in the computation of equilibria, combinatorial games played on graphs, etc ...

But the most durable impact of Game Theory is to all these disciplines where it is used as a modeling tool, a language.

This language is useful in any discipline where the paradigm of methodological individualism and the postulate of rational choice are meaningful. The rational choice model originates in economic theory, where the behavior of consumers and firms is predominantly represented as the maximization of a well defined preference relation (or utility function). Game Theory starts with the rational choice postulate in its most abstract form and explores its logical consequences when several decision makers interact. The only earlier formal analysis of interdependent decision making is the Walrasian model of competitive exchange, where the interaction between the numerous market participants is orchestrated by the price signal and where each market participant is small (i.e., owns few of the resources in the economy).

The reach of Game Theory takes a quantum leap beyond that of Walrasian analysis: it drops the assumption of smallness and is not limited to the exchange and production of commodities. Thus Game Theory is immediately applicable for instance to this branch of Political Science viewing citizens as rational voters and politicians as rational office-holders. It fits just as well the description of the *peer-to-peer* interactions of computers, of animal species competing for reproductive fitness, and so on.

An important feature of the game theoretical insights is to be open ended. The theory does not provide a ready-made template for a context-free analysis of a game. A frequent misconception is that provided the rules of the game and the informational context are specified in enough detail, Game Theory will produce an unambiguous strategic advice to each player, in other words it will teach us a compelling notion of "optimal" rational play. This misguided viewpoint probably originates in the tremendous appeal of Borel and Von Neumann's minimax theorem. In a two-person zero-sum game this result (apparently) identifies an optimal move for each player when he randomly mixes over his deterministic strategic options. The qualification comes from the fact that the players must evaluate the now random outcome of the game by maximizing the expected value of some cardinal utility. This restriction notwithstanding, the minimax theorem all but closes the strategic analysis of two-person zero-sum games, allowing us to compute an optimal strategy for each player under any specification of the game. Repeated attempts at providing a similarly compelling notion of optimal play in general n-person games, or even smaller subclasses of these have proven futile; they bear the names of the two founding fathers themselves, as well as of such luminaries as John Harsanyi and Reinhard Selten.

The great many equilibrium concepts uncovered during 60 years of game theoretical research are now correctly understood to fill a *tool box*, where each concept is well suited to discuss some types of strategic interactions but not others. For instance some of these concepts capture well the exchange of strategic signals in situations of asymmetric information, others focus on the profitable coordination opportunities prompting the formation of coalitions, still others on dynamic environments where past moves may convey a threat or an offer to cooperate, and so on. The point is that a language, in mathematical form or otherwise, is only as good as the questions it can address, and in the case of Game Theory those questions are posed by economists, political scientists, psychologists, sociologists, biologists and even computer scientists. The equilibrium concepts (the tools) are then vindicated by their relevance in those disciplines, and only in that way.

Thus game theorists constantly refine and enrich a language that is spoken in a variety of familiar scientific disciplines, but Game Theory itself is not one of those disciplines, because its object is not a well defined set of questions meaningful to the layman, and because what is and is not a legitimate game theoretical question is ultimately defined by the particular discipline where it is used. We cannot take the object of Game Theory to be any and all interactive decision making problems involving more than one decision maker, as evidently many of these interactions are not governed by the rational choice postulate. But to take its object as comprising only those interactions where the said postulates apply amounts to no definition at all. To take an example, sociology deals with interpersonal relations in the context of social institutions, but for which one of those relations is it useful to invoke a game theoretical model (e.g., a game played between social groups) is the subject of intense debate among experts, not a self-evident call.

The influence of the formal language of Game Theory over the disciplines listed above is in my opinion the ultimate justification of this enterprise, its *raison d'etre*. The vast majority of contemporary game theorists reside academically in one of these disciplines, where they take inspiration for developing new game theoretical ideas. I have not seen for a long time a fresh new idea in the theory that was not inspired by an empirically relevant question from without.

## What do you consider the most neglected contributions in late 20th century game theory?

The normative contributions of Game Theory have a distinct epistemological status, in the usual way in which prescriptive judgments are distinguished from descriptive statements. The archetypal example is the Shapley value for cooperative games (with side payments) discussed in Question 2, an important new idea for fair division. Its success rests on a handful of elegant axiomatic justifications, and the versatility of its applications.

Although the axiomatic component of (cooperative) game theory is a well established sub-field of the discipline, its connection with positive strategic analysis is often overlooked.

The simple observation is that when a game exhibits systematic strategic instability, or symmetrically when the set of its stable outcomes is quite large, normative arguments are the only satisfactory approach to resolve the indeterminacy.

With the terminology introduced in Question 2 above, consider a situation where the core is empty, in other words for any compromise outcome upon which our players may agree, there is some coalition of players with both the means and the interest to renege on the agreement. The simplest example is majority voting when preferences exhibit the famous Condorcet paradox, namely when the binary majority relation is cyclic. Other important instances of core instability are exchange and production games where either consumers have non convex preferences, or firms have non increasing marginal costs. A good example of the latter is the competition of taxis for customers at the airport: the cost of serving one customer is much higher than the marginal cost of the second customer. Core instability in this case results when taxis are in excess supply, because those who end up with only one passenger cannot recover even their operating costs. This cut-throat competition ultimately destroys the coordinating role of markets and may result in an inefficiently low supply of taxis. The remedy is to organize the taxi line at the airport, and impose a uniform fee: a benevolent arbitrator simply eliminates the opportunity to form certain coalitions (a taxi and a pair of customers cutting the line) and imposes a fair outcome. In a voting situation, cycles of the majority relation can similarly be resolved by requesting a qualified majority to change the status quo. The argument generalizes: when the core is empty an arbitrator must limit the power of coalitions, changing the rules of the game so as to bring strategic stability. Such a move must be normatively justified.

In the profit-sharing situation described in Question 2, whether or not the profit opportunities to the coalitions of two out of three partners yield an empty core, strategic analysis does not identify a compelling division of the pie and leaves much room for bargaining among the partners. Again a fair arbitrator is the simplest way out of the crippling indeterminacy.

## What are the most important open problems in game theory and what are the prospects for progress?

This question is by far the hardest, and by way of answer I will only comment on a couple of directions in which my own research is taking me at this point in time.

The field of mechanism design applies Game Theory to social engineering. Its goal is to discover resource allocation procedures meeting three or fewer of the requirements of efficiency (Pareto optimality), fairness and incentive compatibility. In a few cases this approach has produced remarkable mechanisms of great practical significance such as the Gale-Shapley algorithm for bilateral matching or the Vickrey auction, but the overall return per kilo of published research has been poor.

Some elementary theoretical questions still evade a systematical answer. For instance the structure of all strategy-proof mechanisms (see Question 2) is known only in a couple of simple problems like the assignment of one object per agent, or sharing the cost of a binary service (each agent either receives the service or does not).

Unable so far to provide very general structural guidelines for the design of good mechanisms, theoretical work on mechanism design has produced a slew of impossibility results showing for instance that a certain definition of incentive compatibility like strategy-proofness is not compatible with efficiency, or with a simple test of fairness (e.g., horizontal equity: equal treatment of all participants).

Faced with such incompatibilities, the natural next step is to understand the tradeoffs between the various requirements, but this necessitates to measure how far a given mechanism deviates from a certain requirement. We know well how to measure inefficiency, and it is not too hard to come up with a measure of unfairness for most tests of equity. But how to measure the deviation of a given mechanism from strategy-proofness?

# 13

# Rohit Parikh

Distinguished Professor
Brooklyn College of CUNY and CUNY Graduate Center
City University of New York, USA

---

I want to start by quoting something which Ariel Rubinstein has said:

> There are many similarities between logic and game theory. Whereas logic is the study of truth and inferences, game theory is the study of strategic considerations. Logic is motivated by the way in which we use the notions of truth and inferences in daily life while game theory is motivated by the strategic considerations we use in daily life.

Indeed, there are many connections between logic and games. Logic studies reason, whereas game theory studies rationality, but the two are connected. Even the 'many agents' aspect of game theory has its parallel in the Abelard-Eloise games used to understand the semantics of first-order logic, or their formal counterpart, Ehrenfeucht games [6], or most recently, the game theoretic versions of IF-logic [15, 38].

## Why were you initially drawn to game theory?

I was initially drawn to game theory because of an obvious connection (but not identification) between being logical and being rational. Neither implies the other, but there is a correlation. Also, there were issues in which I was interested where a game-theoretic approach made sense.

At a practical level, my interest began because of Aumann's "Agreeing to disagree" paper [2]. There was a sequel to that paper by Geanakoplos and Polemarchakis [9] which interested me and my then student Pawel Krasucki. G&P had proved their result for two players, and the many player case (with pairwise communication) was open. Krasucki and I proved a strong extension to the G&P paper which appeared in *J. Economic Theory* [32], and which was followed by other papers by Krasucki alone. We considered *convex* set functions i.e., functions from the power set of a space $W$ to the reals, and satisfying a somewhat stronger requirement than Savage's *sure thing principle*. This requirement is that if $X$ and $Y$ are disjoint, and $f(X) < f(Y)$, then $f(X) < f(X \cup Y) < f(Y)$, i.e., $f(X \cup Y)$ lies properly between $f(X)$ and $f(Y)$. Of course if $f(X) = f(Y)$ then $f(X \cup Y) = f(X)$. Both probabilities and expected values of random variables are convex functions in this sense. We show that if a finite number of agents repeatedly exchange (in pairs) their latest value of such a convex function, and update appropriately, then the values become equal, provided that no agent is left out of the chain of communication. It is not necessary that every agent speak to or hear directly from every other agent.

But my work with Krasucki, while inspired by the work of game theorists, was not in game theory proper. My interest in game theory proper became serious when my friend Bud Mishra invited me to co-direct the dissertation [11] of Amy Greenwald at NYU and I attended some of the classes in the game theory course he was teaching. The three of us also worked on the Santa Fe bar problem [1, 10]. For some time after that I was a regular attendee at the theory seminar at NYU's Starr Center in economics; I even sat in on Benoit's course on game theory. So I enjoyed contacts with the faculty there as well as with visiting faculty. Eventually I taught a course on game theory myself at the CUNY Graduate Center, using the book by Osborne and Rubinstein [20].

A different direction for approaching game theory came about as a result of my developing the logic of games. During the late 70's and early 80's, I had been working on dynamic logic, a tool invented by Vaughan Pratt [40, 13] for proving the correctness of computer programs. Along with Dexter Kozen (then at IBM, now at Cornell), I had already proved a completeness result [18] for propositional dynamic logic (PDL). Then I suddenly realized that much of the logical apparatus developed for studying programs applied also to games. I showed [30] how a completeness result for a propositional logic of games could be proved, and applied the techniques to provide a formal proof of the correctness of the Banach-Knaster last-diminisher procedure for fairly dividing a cake. This correctness consists in the fact that every player has a winning strategy for obtaining at least $1/n$ of the cake by his own personal measure.

That work was followed up by Marc Pauly, then doing a dissertation at Amsterdam [39], and he used a similar intuition to develop a logic for coalitions. I was a member of Pauly's dissertation committee, and this fact brought me in close contact with the Amsterdam group, especially Pauly's advisor, Johan van Benthem.

Thus three different streams, one associated with Aumann's agreeing to disagree paper, one associated with Greenwald's thesis, and one with the logic of games, brought me into serious contact with game theory.

When I was working with Bud Mishra and Amy Greenwald, I very much wanted to meet Steven Brams, in the department of Political Science at NYU, but this only happened later, when I taught Logic in the Philosophy department at NYU. I was very interested in voting theory and fair division, both of which were areas of Brams' expertise, and I eventually ended up publishing in both areas, an expository article [35] in the CUNY Graduate Center Philosophy program's journal *Aurora*, edited by Mariya Gluzman, and later, more technical papers with Eric Pacuit, Samir Chopra, and Samer Salame in various venues [25, 26, 5].

**What example(s) from your work (or the work of others) illustrates the use of game theory for foundational studies and/or applications?**

**What is the proper role of game theory in relation to other disciplines?**

I am going to conflate these two questions.

I think game theory is a very elegant but *preliminary* approach to understanding human behaviour in the context of competition and co-operation. While these activities occur naturally in the marketplace, they also occur elsewhere, and game-theoretic methods have even given insight into biology.

**Logic via Games:**

A connection between the two fields with games as explaining logic is quite old. In a common interpretation of first-order logic, the two quantifiers, existential and universal, are interpreted as moves by two players, Eloise and Abelard respectively. In the usual interpretation of such games, the given formula is true in the given model if Eloise has a winning strategy and is false if Abelard does. Suppose for instance that our universe is the natural numbers and we consider the (true) formula $(\forall x)(\exists y)(x < y)$. Then in the corresponding game, Abelard ($\forall$) first chooses the $x$ and then knowing the value of $x$, Eloise ($\exists$) chooses a $y$ which will make $x < y$ true. Clearly she can do this by taking $y = x + 1$. On the other hand for the formula $(\exists y)(\forall x)(x < y)$, Eloise has to choose $y$ *before* Abelard makes his move and naturally she cannot guarantee that $x < y$. The formula $(\exists y)(\forall x)(x < y)$ is false. Such games are determined for first-order logic, so that one player always has a winning strategy. In more recent logics like IF-logic whose proponents are Hintikka and Sandu [15], it may happen that neither Eloise nor Abelard has a winning strategy so that the law of excluded middle fails. If neither player has a winning strategy then we could say that the corresponding formula is not true (Eloise cannot guarantee a win), but neither is it false (since Abelard also has no winning strategy).

In *finite information logic* investigated in [38], Eloise only has a finite amount of information about Abelard's moves. It turns out that this leads to an interesting, decidable logic. This logic may be appropriate for certain social applications, as many personal and social decisions are only made on the basis of a finite amount of information.

In the early 70's, some colleagues and I provided a common game theoretic semantics for first-order logic, intuitionistic logic and Ehrenfuecht's *-logic [7, 8] using what we called D-structures. That work was published in abstracts [28, 29, 16] in the early 70's, but as a paper [33] only very recently.

**Games and Language:**

Another field where the importance of game theory is known mainly to the cognoscenti (by definition?) is linguistics. Pragmatics is of course a very old branch of linguistics, and pragmatic considerations occur in both Peirce's and Ramsey's work. However, these pragmatic considerations become much more game-theoretic with Grice. His notion of implicature [12] makes it clear that an utterance can be seen, and indeed must be seen, not merely as a conveying of information, but also as a move in a game.

Thus when A says to B, "My car is out of gas," and B responds, "There is a gas station around the corner," B has *implicated*, but not *said*, that the gas station is open. That this is an implicature and not part of the meaning is revealed by the fact that B could then go on to say, "but it is not open today." An implicature is *cancellable*.

Grice, although he does not use game-theoretic terminology, regards a conversation as a co-operative game between a speaker and a hearer. If B were to make his statement about the gas station even though he knows it is closed, he *would* have made a *true* statement but which would result in A, who needs gas, making a fruitless trip to the site of the gas station. Thus if A believes that B is playing a co-operative game, A can conclude that the gas station must be open, or at least that it is as far as B knows. This is how an implicature can go *beyond* the bare meaning of the sentence. Prashant Parikh [27] develops such a game-theoretic account explicitly, analyzing the resolution of semantic ambiguities using Nash equilibria of communication.

The paper [31] shows that the problem of vague predicates can be solved if, instead of asking, *What statements involving vague predicates are true?*, which leads us into a paradox (the *Sorites paradox*), we ask instead, *What function do vague predicates fulfill in human communication?* It turns out that in certain contexts, the use of vague predicates allows us to act faster and more efficiently. Someone looking for a book will be *helped* by the statement that the book is blue, even though the word *blue* is vague, and the speaker and the listener may have different notions of what objects are blue. All that is needed is that the correlation between the two extensions of *blue*, the speaker's extension $blue_s$ and the listener's extension $blue_\ell$ be positive. It would of course be unrealistic to demand that the extensions be *exactly* the same, as color perception depends on chemical processes going on in our eyes which do not repeat themselves.

## What do you consider the most neglected topics and/or contributions in late 20th century game theory?

Needless to say, the fact that Homo Sapiens is not Homo Economicus is being more and more accepted these days, starting with paradoxes like that of Allais, and more recently with the work of Kahneman, Slovic and Tversky [17], and the counter-intuitive (to game theorists) results of studies in the ultimatum game across cultures. Thus the field of behavioural economics is flourishing. But there are also other issues which game theory needs to address.

I would say that the two most important issues which could use more emphasis are knowledge, and logical or algorithmic structure. I will illustrate the first with an example.

## The Role of Knowledge:

Near Wassenaar, Netherlands, where I stayed for six weeks, there is a regular bus service and the bus stops consist of a small shelter with a bench inside. The sides of the shelter are transparent, but there are also large advertisements in these shelters. However, these advertisements are there on only *one* side of the shelter, and the side from which a bus would approach is left transparent. This enables the driver of the bus to see if there is someone sitting on the bench, waiting for a bus. It also allows the passengers to see if there is a bus coming. Thus there is *information transfer* which would not take place if both sides had advertisements on them, thereby making them opaque.

Here there is only a small element of game-theoretic strategizing, and the transmission of information is the crucial element. Designing proper social algorithms should involve taking knowledge transmission into account. It is not enough that a building *has* restrooms, it also needs signs saying *where* they are.

I shall give another example of an issue which is both algorithmic and knowledge theoretic. When I buy a train ticket, then the ticket, as document, is proof that I have paid the fare, and can ride the train. If it is a traditional plane ticket, it is also such a proof, and typically not transferable to another person. So it serves a knowledge theoretic function, namely that of serving as certificate for a certain right.

However, the traditional ticket also served another function. When surrendered to my employer it constituted proof, not only of the fact that I had travelled but also of the fact that I had not claimed the money from another source. And that is because there was only *one* plane ticket per journey. This is the reason why employees are asked for the original tickets and not a photocopy.

But now, with e-tickets, there is a problem. There is no problem between me and the airline, they can use an e-ticket as proof that I paid, and I can use it as proof of my right to travel.

With an e-ticket, I can make as many copies as I like! Thus I could claim re-imbursement from my principal employer, and from another one if I have one, and also deduct it from my taxes! Thus for *certain* purposes, the e-ticket has different algorithmic properties from the traditional ticket.

Thus the e-ticket, which has many advantages, also has a disadvantage! Employers sometimes deal with this by also asking for the boarding pass, of which there is only one. If there were also an e-pass, then this method would cease to work!

Similar issues arise with an election where we want each voter to vote only once. Voters' hands may be marked with a stamp when they vote, and the ink had better be indelible!

A more serious example of the importance of knowledge is discussed in [24]. A doctor whose neighbour is ill has no obligation to treat him. But if she is informed that he is ill, then she does have such an obligation. You might have said, *What? a doctor whose neighbour is ill has no obligation to treat him?* But you were taking the knowledge condition for granted! The notorious Valerie Plame case which brought much embarrassment to the Bush administration is also a knowledge problem – *Who knew without proper authorization that Plame was a CIA agent?* – and is discussed in [22, 21].

**Algorithmic Issues:**

A paradigm example in game theory is a one-shot game. We know that in the prisoner's dilemma, defecting is the dominant strategy in a one shot game, and even for repeated games, by backward induction, defecting all the way is the indicated Nash equilibrium. The only condition needed (according to one view) is common knowledge of rationality and common knowledge of the end of the game.

But more complex procedures than one shot or repeated games do occur in ordinary life and need to be considered. For instance, the standard last-diminisher procedure for cake cutting takes $O(n^2)$ steps. It takes $n$ steps for one player to take his/her slice and leave, and $n-1$ steps for the next one and so on, so that we end up with $O(n^2)$ steps for the whole procedure.

Here the entire argument is algorithmic and quite reminiscent of considerations in the computer science sub-area, *Analysis of Algorithms*. Surely, a good analysis of a social algorithm must involve not only the preferences of the players, but what knowledge they have and when, and what the logical structure is of the procedure which they are following. The running time of the algorithm can also matter. If a phone company were to write down names and phone numbers in an unsorted list, then the amount of time taken to find a phone number would be proportional to the number $n$ of subscribers. But when they are in alphabetical order, then the amount of time is the much smaller logarithm of $n$.

There are cases when even an algorithm which is purely individual calls on methods from the analysis of algorithms. For instance in [37] Parida, Pratt and I look at the quandary of someone who has just washed socks and now wants to match them up in pairs. The difficulty is that because color perception is vague, color matching is intransitive. Thus it may happen that (the original) pair of socks $s, s'$ matches the pair $s_1, s'_1$ which matches $s_2, s'_2$ but that the socks $s, s_2$ do not match. If we then put together socks $s_1, s_2$, which appear to match, then there is no way to match up the remaining four socks. Naturally the problem is much harder with say 50 socks! This looks like a possibly NP-complete problem, but in fact there is a quadratic time algorithm.

Of course, that paper was written largely for amusement, but there are devices like file cabinets or staplers, and structures like queues at bus stops whose role is primarily algorithmic in a social context.

Perhaps it is only in terms of my own interests that I consider the issues of knowledge and algorithmic structure to be so important. But I have argued that they *are* relevant. And that brings me to my own pet project of social software.

**Social Software:**

Over a ten year period starting with the mid 90's I became convinced that all these topics – game theory, economic design, voting theory, etc. – belonged to a common area which I called *Social Software* [34, 36]. This insight was not entirely new, and indeed I recently learned that Peyton Young, who had been at the CUNY Graduate Center before me had also taught there a course on Mathematical Methods in the Social Sciences. But Young was not a computer scientist, and anyway, the areas of Computer Science, *Analysis of Algorithms* and *Logic of Programs*, which are relevant to Social Software barely existed when he was at CUNY.

The main insight in Social Software is that society consists largely of people carrying out various algorithms (or software) to achieve various ends. Thus I might take a taxi, then a plane, and then a limousine to arrive at a hotel where I will stay when a certain conference is taking place.

In this case the algorithm consists of three definite steps. But one can also have recursive steps in ordinary algorithms, like *whip the cream until it thickens* in some cookbook recipe. This step does not say, *whip ten times*, or even *whip for five minutes* but *whip until the cream thickens*. Thus a test of whether the cream has thickened yet, performed repeatedly, is involved. Perhaps president Bush is also carrying out such a recursive algorithm consisting of the recursive step, *Stay in Iraq until it becomes a viable democracy.*

Naturally, there are non-termination dangers in such recursive steps. If I am beating skim milk rather than cream, it might never thicken, and perhaps (as so many fear) president Bush's algorithm for Iraq will also never terminate.

Normally, we do not perform algorithms in isolation. For me to take a taxi or a plane, these services have to be already provided by society. Moreover, some algorithms, like moving a piano, or giving a lecture or holding an election, involves many agents.

A fairly comprehensive survey of recent work in Social Software is in my joint paper [21] (with Pacuit) appearing in the volume *Interactive Computation.*

## What are the most important open problems in game theory and what are the prospects for progress?

I think that game theory will have to split into two different areas. Classical or mathematical game theory is by now too beautiful an area to be abandoned completely. So it will survive rather like a branch of pure mathematics, just as number theory does; mostly for its intrinsic interest, but occasionally also yielding insights, as number theory has done for cryptography.

On the other hand, to the extent that game theory is the study of human behaviour in groups, it will have to ally itself more and more with fields like psychology and computer science. We are all aware that the predictions made by classical game theory are not always borne out. Culture and ingrained habits of action play a large role.

I also believe that at least *some* people interested in games should take a serious interest in the language games of Ludwig Wittgenstein [41]. When he spoke of *language games*, Wittgenstein was emphasizing what I would call the algorithmic element in language, and by implication also in philosophy. But the applications of his ideas have a wider scope. My view is that Game theory sits inside life and language, and these connections will, of necessity, need to be looked at from time to time.

## 13.1 REFERENCES

[1] W.B. Arthur, "Inductive reasoning and bounded rationality," *Complexity in Economic Theory*, **84**, 406–411, 1994.

[2] R. Aumann, "Agreeing to disagree," *Annals of Statistics* **4** 1976, 1236–1239.

[3] S. Brams and A. Taylor, *Fair division*, Cambridge U. Press, 1996.

[4] S. Brams and P. Fishburn, "Voting procedures," in *Handbook of Social Choice and Welfare*, edited by Arrow et al, Elsevier, to appear.

[5] S. Chopra, E. Pacuit and R. Parikh, "Knowledge–Theoretic Properties of Strategic Voting," in José Júlio Alferes, João Leite editors. *Logics in Artificial Intelligence: 9th European Conference, JELIA 2004*. Lecture Notes in Computer Science. pgs. 18–30.

[6] A. Ehrenfeucht, "An application of games to the completeness problem for formalized theories," *Fundamenta Mathematica*, **49** (1961) 129–141.

[7] A. Ehrenfeucht, "Logic without Iterations" *Proceedings of the Tarski Symposium* (1974) pp. 265–268.

[8] A. Ehrenfeucht, J. Geiser, C. E. Gordon and D. H. J. de Jongh. "A semantics for non iterated local observation," Preprint 1971.

[9] J. Geanakoplos and H. Polemarchakis, "We can't disagree forever," *J. Economic Theory* **28** (1982) 192–200.

[10] A. Greenwald, B. Mishra and R. Parikh, "The Santa Fe bar problem revisited" presented at the Stony Brook workshop on Game theory, Summer 1998.

[11] A. Greenwald, *Learning to Play Network Games*, Ph.D.-dissertation, NYU 1997.

[12] P. Grice, *Studies in the Way of Words*, Harvard University Press, 1989.

[13] D. Harel, D. Kozen and J. Tiuryn, *Dynamic Logic*, MIT Press, 2002.

[14] J. Hintikka, *Knowledge and Belief*, Cornell U. press 1962.

[15] J. Hintikka and G. Sandu, "A Revolution in Logic," *Nordic Journal of Philosophical Logic*, 1, 1996, pp. 169–183.

[16] D.H.J.de Jongh, N. Goodman and R. Parikh "On Regular *-structures with Classical Theories," *J. Symb. Logic* 37 (1972) 777.

[17] D. Kahneman, P. Slovic, and A. Tversky, *Judgment under Uncertainty: Heuristics and Biases*, Cambridge University Press, 1982.

[18] D. Kozen and R. Parikh, "An Elementary Completeness Proof for PDL," *Theor. Comp. Sci* 14 (1981) 113–118.

[19] D. Lewis, *Convention: A Philosophical Study*, Harvard U. press 1969.

[20] M. Osborne and A. Rubinstein, *A Course in Game Theory*, MIT Press 1994.

[21] E. Pacuit and R. Parikh, "Social Interaction, Knowledge, and Social Software," in *Interactive Computation, The New Paradigm*, edited by Goldin, Smolka and Wegner, Springer 2006, 441–462.

[22] E. Pacuit and R. Parikh, "Reasoning about Communication Graphs," Presented at Augustus de Morgan Workshop: Interactive Logic: Games and Social Software, 2006 (ADMW 2006). To appear in the *proceedings of ADMW*.

[23] E. Pacuit, *Topics in Social Software: Information in Strategic Situations*, doctoral dissertation, CUNY Graduate Center, 2005.

[24] E. Pacuit, R. Parikh and E. Cogan "The logic of knowledge based obligation," *Synthese*, **149** (2006), 311–341. Appeared also in *Knowledge, Rationality and Action*.

[25] E. Pacuit, R. Parikh and S. Salame, "Some Results on Adjusted Winner," presented at International Game Theory Conference (Stony Brook University, 2005).

[26] E. Pacuit and R. Parikh, "Knowledge Considerations in Strategic Voting," presented at a Workshop on Voting Theory and Preference Modelling, Unversity of Paris Dauphine (October 25–28, 2006).

[27] P. Parikh, "Communication, Meaning, and Interpretation," *Linguistics and Philosophy*, **23**(2) 2000, 185–212.

[28] R. Parikh, "D-Structures and their Semantics," (abstract) *Notices of the American Math. Society* **19** (1972) A329.

[29] R. Parikh and J. Mayberry, "D-structures and *-structures," *Notices of the American Math. Society* **19** (1972) A454.

[30] R. Parikh, "The Logic of games and its applications," *Annals of Discrete Math.*, 24 (1985) 111–140.

[31] R. Parikh, "Vagueness and Utility: the Semantics of Common Nouns," in *Linguistics and Philosophy* **17** 1994, 521–35.

[32] R. Parikh and P. Krasucki, "Communication, Consensus and Knowledge," *J. Economic Theory* **52** (1990) pp. 178–189.

[33] R. Parikh, "*D*-Structures and their semantics," appeared in a volume dedicated to Johan van Benthem, University of Amsterdam, August '99.

[34] R. Parikh, "Language as social software" in *Future Pasts: the Analytic Tradition in Twentieth Century Philosophy*, Ed. J. Floyd and S. Shieh, Oxford U. Press, 2001, 339–350.

[35] R. Parikh, "Some thoughts on election procedures," in *Aurora*, Jan 2001. (Doctoral program in Philosophy, CUNY Graduate center).

[36] R. Parikh, "Social Software," *Synthese*, **132**, Sep 2002, 187–211.

[37] R. Parikh, L. Parida and V. Pratt, "Sock Sorting," appeared in a volume dedicated to Johan van Benthem, University of Amsterdam, August 99, reprinted in *Logic J. of IGPL*, vol 9 (2001).

[38] R. Parikh and J. Väänänen, "Finite information logic," *Annals of Pure and Applied Logic*, **134**, 2005, 83–93.

[39] M. Pauly, *Logics for Social Software*, doctoral dissertation, Amsterdam University, 2001.

[40] V. Pratt, "Semantical considerations in Floyd-Hoare Logic," in *Proc. 17th Annual IEEE Symposium on Foundations of Computer Science*, (1976) 109–121.

[41] L. Wittgenstein, *Philosophical Investigations*, Translated by G.E.M. Anscombe, Basil Blackwell 1953.

# 14
# Ariel Rubinstein

School of Economics, Tel Aviv University, Israel
Department of Economics, New York University, USA

---

Why were you initially drawn to game theory?

I could answer this question by simply saying that I studied at the Hebrew University of Jerusalem which was home to many of the giants of game theory and related fields: Yisrael Aumann, Arie Dvoretzky, Bezalel Peleg, Menachem Yaari, Michael Rabin and Amos Tversky. So what else would you expect? But that would only be a partial answer.

I could say that it is the ingenious name given to the field – game theory – which attracted me. I doubt if I would have chosen a field called "Theory of rationality and decision making in interactive economic situations." But in fact my first encounter with game theory was a disappointment. In my second undergraduate year (1972–3), I tried out a course given by the Mathematics Department entitled *Introduction to Game Theory*. I remember that the lecture hall was full and the lecturer very enthusiastic. He started the course with some abstract theorems on convexity. I left before the end of the first class.

I could also say that I chose game theory because I wanted to improve my strategic skills for the crusades on which I was hoping to embark in the future or to improve my negotiating abilities in the open-air markets of Jerusalem. But that wouldn't be right either. I have never thought of game theory as being useful in a practical sense. In fact, I was quite shocked in 1987 when I discovered for the first time that some of my fellow economic theorists believed that a model should be confirmed in the laboratory or using real empirical data.

The seeds of my interest in game theory were planted during my undergraduate studies in mathematics at the Hebrew University. While I admired the intellectual beauty of the material, I had a vague notion that, despite its abstractness, mathematics had some connection to real life. So I tried to superimpose the mathematical models onto the subject that occupied my thoughts both then and now: the realm of human interaction. Somewhere between mathematics and the study of human interaction, game theory awaited me.

## What example(s) from your work (or the work of others) illustrates the use of game theory for foundational studies and/or applications?

Implicit in this question is the idea that game theory can and probably should be evaluated according to its usefulness. The phrase "the use of game theory" which appears in the question sounds analogous to "the use of physics in the design of rockets" or "the use of biology in the identification of genetic diseases." In my opinion, it isn't analogous.

The discussion of the usefulness of game theory (or for that matter economics in general) is charged with emotion and subject to misunderstandings. The everyday terminology of game theory attracts people's attention but for the wrong reason. Human beings are eager to find professional solutions to problems they tackle. They look for techniques and ideas to improve their strategic skills as if they were weight training to build up their athletic skills. In my thirty years in the profession I have not encountered a single case in which game theory has provided a solution to a real problem and have not found any evidence that it has the ability to improve strategic thinking.

An article I read in the Israeli newspaper *Haaretz* while writing this essay demonstrates the public's confusion about game theory. A former senior politician was writing about the current tension between Iran and Israel. He claims that game theory is already able to explain the interactions between two rational players. He also states that according to game theory an irrational player has an advantage over a rational one. (In my opinion, this is a myth promoted by hardliners who want to persuade rational people to act "tough.") But then he claims that at the moment no one knows how to analyze a game between two irrational players. He goes on to assume that the President of Iran is irrational and that the Israeli government has recently adopted an irrational strategy by appointing one of the most controversial politicians as the Minister in charge of dealing with strategic threats. This leads him to call on game theorists, and Robert Aumann in particular, to "save us."

This person obviously takes game theorists seriously when they claim that game theory is useful. This claim is often made and not only in NSF proposals. Almost any survey of game theory starts with a sentence like "Game theory is useful in a wide spectrum of fields—from Botany, Zoology and Medicine through Economics, Management, Computer Science and Politics to History and Biblical Studies." However, the fact that the "prisoners' dilemma" is mentioned in a text does not make it an application of game theory. And the fact that game theorists are involved in a discussion does not make it an application of game theory.

Let us recall that game theorists and economists are in the end only human. Paradoxically, we assume that all agents in the world are selfish and manipulative and act to advance their own interests, but somehow we are not used to thinking of ourselves in this way when we assess the usefulness of our own models.

I believe that one of society's goals should be the pursuit of knowledge and scholarship for their own sake. For me, game theory is an investigation of the ways in which human beings think in interactive situations. Even if game theory is of no practical use, it still has value as part of our continuing investigation of the mind.

## What is the proper role of game theory in relation to other disciplines?

How would one answer the following question: "What is the proper role of logic in relation to other disciplines?" I would argue that if the word "logic" were replaced with "game theory" the answer to this question would be the same.

There are many similarities between logic and game theory. Whereas logic is the study of truth and inference, game theory is the study of strategic considerations. Logic is motivated by the way in which we use the notions of truth and inferences in daily life while game theory is motivated by the strategic considerations we use in daily life. I doubt whether there is a type of logic which is "right"; in the same way, there is no one type of game theory that is "right". The standard rules of logic have a unique status; similarly, rationality has a unique status in economics. Both logic and game theory are analyzed using formal models. Logic does not induce people to think logically just as game theory does not induce them to think strategically. So what is the role of logic or of game theory in relation to other disciplines? The answer is simply that both provide a limited set of ideas and tools for use in those other disciplines.

Let me demonstrate the proper role of game theory in relation to other fields using a project I am involved in together with Kobi Glazer (see for example, Glazer and Rubinstein (2006)). The project involves research into pragmatics, a branch of linguistics that explores the rules that determine how people interpret an utterance made in a conversation beyond its literal content. We are interested in persuasion situations where an informed party wishes to persuade an undecided and uninformed party to adopt his position. The difference between this situation and a regular conversation lies in the interests of the two parties. In a conversation there is an underlying assumption that the two parties have common interests while the parties involved in a persuasion situation have at least a partial conflict of interest. The speaker wants to persuade the listener to believe what he says while the listener wishes to be persuaded only under certain circumstances.

We noticed that a persuasion situation is subject to different rules of pragmatics than a conversation. For example, assume that you are discussing the chances of each of two candidates – $A$ and $B$ – in upcoming elections and that the electorate consists of nine voters. Assume that the speaker knows the positions of all the voters but due to time constraints can only present the views of three of the nine (who are enumerated as $c_1$ to $c_9$). The speaker claims that candidate $A$ will win and presents evidence that $c_1$, $c_4$ and $c_8$ support $A$. If it is a friendly conversation and the speaker's interests are similar to yours, then you are likely to think that he has selected three people who represent the view of the majority. Thus, you are likely to conclude that $A$ will win the election. If, on the other hand, it is clear that the other person is trying to persuade you that $A$ will win, regardless of whether this is true or not, you will doubt his claim since you suspect that he has intentionally selected three supporters of $A$ and that $c_2$, $c_3$, $c_5$, $c_6$ and $c_7$ weren't mentioned, even though they appear before $c_8$ in a list, because they support $B$.

In this project, we attempt to provide an explanation for this sort of pragmatic phenomenon using an "economic/game theoretic approach." We assume that pragmatic rules of persuasion are determined by an imaginary designer prior to the conversation. The rules of pragmatics determine the "game" played between the speaker and the listener in such a situation. We assume that the designer wishes to maximize the probability that the listener will make the "right" decision on the basis of the information provided to him by the self-interested speaker subject to the constraints on the amount of information that can be conveyed. In such a model one can show that according to optimal persuasion rules the presentation of evidence regarding certain combinations of voters is persuasive while others, which include the same number of voters, are not.

Our investigation applies economic/game theoretic principles to pragmatics. I don't know whether people in other fields have thought along similar lines, but I am quite certain that stating these ideas clearly requires the sort of formal and conceptual tools that have been developed and used in economics and game theory. But I am also aware of the fact that the assessment of the results of such an approach requires the expertise possessed by philosophers and linguists rather than game theorists. Game theory's tools can produce possible explanations but cannot evaluate them.

## What do you consider the most neglected topics and/or contributions in late 20th century game theory?

I do not feel comfortable with the term "neglected topic" which seems to imply that there are areas which game theory should be investigating but isn't. Thus, I will simply list five topics in game theory that represent significant contributions during the late $20^{th}$ century:

1. **The Interpretation of Game Theory**: My impression is that more and more game theorists are struggling with the interpretation of game theory. Is it a theory as that term is understood in the sciences or is it a collection of fables (see Rubinstein (2006))? This is not a question that can be settled one way or another but the discussion and clear statement of the issues are crucial.

2. **Behavioral Game Theory**: In both economics and game theory, behavior is defined as rational if it can be described as an attempt to advance a well-defined goal. In applications of game theory and economics, rationality is usually defined more narrowly so that the goal is physical and "rational," such as the probability of survival or the level of

consumption. The application of game theory which also relates to goals has become increasingly common during the last two decades under the rubric of Behavioral Economics. Although this shift has not affected abstract game theory, which is indifferent to the content of the preferences, it is part of a major trend in economics in general and has had a major impact on the way in which game theory is applied.

3. **Models of Bounded Rationality**: Little work has been done to develop game theoretic models in which rationality is replaced with alternative choice procedures. The development of theories of interaction between agents who don't behave according to the rational man paradigm requires a major change in the solution concepts and not just in the payoff functions (which is all that is needed in order to include psychological elements within the model). One example of such a model is Osborne and Rubinstein (1998). In that model, we assume that each player constructs beliefs about the consequences of his actions on the basis of past experience. Thus, he attaches to each action the consequence which was observed when the action was taken on previous occasions. An S-1 equilibrium consists of a distribution of actions among each of the players such that the probability assigned to a particular action being played by a particular player is the probability that the player will consider that action to be optimal given his random sampling of past experience. This solution concept has some desirable properties. For example, the repetition of an action can affect the solution and a dominated action can still be played with positive probability.

4. **Experimental Game Theory**: We have seen significant development in this area though I am not happy with the direction it has taken. Researchers in this field insist on experiments being carried out in laboratories and using monetary rewards. I feel this to be unnecessary and simply intended to create barriers of entry. In addition, the field is characterized by small and unrepresentative samples and hastily-drawn conclusions and there is no widespread practice of replicating experiments.

5. **Neuro Game Theory**: This is a new trend in game theory in which researchers attempt to explain behavior by observing brain activity. Unfortunately, this line of research has

gotten ahead of itself. At this stage, the conclusions drawn are wildly speculative. But, of course, I cannot rule out the possibility that significant progress in the understanding of brain functioning in general will some day provide interesting ideas for Game Theory as well.

## What are the most important open problems in game theory and what are the prospects for progress?

The term "open problems" may be appropriate for a field like mathematics in which the problems are usually clear-cut and simply waiting for a genius to solve them but it isn't relevant for game theory whose main goal is to formulate and clearly state problems. In any case, I would like to refrain from simply listing ideas for future research in game theory and that is for two reasons: (i) If the idea is original then I would use it myself, ... and (ii) If my suggestions lead others to develop interesting models, then they might feel obliged to give me credit that in fact I wouldn't deserve. A worthy achievement in game theory does not involve declaring some vague goal or inventing some catchphrase but rather building a simple but rich model that enables one to derive interesting results.

## References

Glazer, Jacob and Ariel Rubinstein (2006). "On the Pragmatics of Persuasion: A Game Theoretical Approach," *Theoretical Economics*, forthcoming.

Osborne, Martin and Ariel Rubinstein (1998). "Games with Procedurally Rational Players," *American Economic Review 88*, 834–847.

Rubinstein, Ariel (2006). "Dilemmas of An Economic Theorist," *Econometrica*, 74, 865–883.

# 15

# Larry Samuelson

Professor of Economics
University of Wisconsin, USA

## Comments on Game Theory

## 15.1 Introduction[1]

From its formal beginnings in von Neumann and Morgenstern [49] and Nash [39], it was clear that game theory and economics were a natural match—a sentiment reflected in von Neumann and Morgenstern's title. Game theory nonetheless remained an active but somewhat isolated field within economics for some time, until the strategic revolution of the 1980s carried its influence throughout the discipline. Game theory is now part of the standard graduate school training and the standard economics tool kit, whether one is a "game theorist" or not, typically being used without any special notice. Indeed, economics has done a lively export business in game theory, spreading the latter throughout the social sciences and into the natural sciences.

I was one of the economists swept up by the strategic revolution of the 1980s. Game theory was largely absent from the core microeconomics courses in my graduate program, being presented instead in a separate topics course that was considered both optional and somewhat exotic, just as was a similar course on growth theory or a course on uncertainty and information. My early work reflected my graduate education in centering around the analysis of competitive markets. But I was captivated by the explosion of

---

[1] I thank the National Science Foundation (SES-0241506 and SES-0549946) for financial support.

work in the 1980s applying game theory to new economic problems.[2] Foremost among these was the game-theoretic study of bargaining, pioneered by Rubinstein [43] and promising the ability to make progress on an issue that had hitherto been regarded as simply outside the purview of economic theory. I was quick to join the study of bargaining under incomplete information. It soon became apparent, however, that further progress on economic applications required a deeper understanding of the foundations of game theory. This directed my attention to more basic issues, including the implications of rationality, the relationship between normal-form and extensive-form games, and the study of evolutionary game theory.

What has game theory accomplished in the course of its development into a standard economics tool? What are its successes and failures, and how do we evaluate them? What remains to be done? This essay offers some perspective on these questions and on game theory more generally.

## 15.2 Who is Game Theory About?

Game theory is not monolithic. Games have been used to capture different ideas and for different purposes. Making these differences clear, while noting that different applications may call for different techniques, takes us a long way toward understanding game theory. It is helpful to start with the players. Who is game theory about?

### 15.2.1 The Classical View: Perfectly Rational Players

One possibility is that game theory is meant for an idealized world of perfect rational players. Much of the interest in pursing this view of game theory comes from the fact that it is not completely clear what "perfectly rational" means. For a decision theorist, a good point of departure would be that a perfectly rational agent is a Bayesian expected utility maximizer. One could then work backward to uncover the foundations of this view and work forward to explore its implications, but no difficulties would arise as long as the environment in which the agent is making decisions is suitably well specified. The difficulty in game theory is that this

---

[2] See Section 15.3.3 for some examples.

environment is *not* obviously well specified. The rational thing for player 1 to do depends upon what (the rational) player 2 does, which in turn depends upon what player 1 does, and so on, introducing a self-reference that calls the very notion of rationality into question. This approach to game theory can be viewed as a philosophical inquiry into the meaning of rationality. The result has been some beautiful theory (see, for example, the survey by Dekel and Gul [8]), but with aspects that sometimes – as in the importance of the distinction between probability zero and impossibility – rival the angels on the heads of the philosophers' pins.

This focus on perfect rationality brings with it two features. First, the game itself is taken to be a literal description of the strategic interaction of interest. Any opportunities for the players to make commitments, to communicate, to make use of outside information, to avail themselves of outside opportunities, and so on, should be contained in the model. Kohlberg and Mertens [27, p. 1005] offer one of the clearest statements: "We adhere to the classical point of view that the game under consideration fully describes the real situation—that any (pre)commitment possibilities, any repetitive aspect, any probabilities of error, or any possibility of jointly observing some random event, have already been modeled in the game tree."

Second, the appropriate concept of equilibrium, and its implications for behavior in any particular circumstance, should be an implication of the theory. All of the information required to determine equilibrium behavior is included in the specification of players, strategies, and payoffs, interpreted through the assumption of perfect rationality. Harsanyi and Selten [22] present the most fully developed implementation of this point of view, offering a theory of rational play capable of identifying a unique equilibrium for any game.

## 15.2.2 The Humanistic View: People

An alternative point of view is that game theory is meant to be about people, the same sort of people whose behavior is studied by the rest of economics. If there is anything we know about such people, it is that they are most definitely *not* always perfectly rational. One need only walk down the self-help aisle of a bookstore, or lapse into a moment's introspection – who has a completely-formed prior over the future course of their life? – to believe in a good dose of irrationality. At the same time, this constitutes no contradiction with the premise that people often do a good job of making their decisions, and can often be usefully described as maximizing. However, it does suggest that the foundations of our models must come from other sources than theories about rationality.

We refer to this as the "humanistic" view of game theory. This focus on people raises three issues. First, how are we to model the behavior of people, if we cannot religiously insist on rationality? The resulting literature on bounded rationality has grown alongside game theory, culminating in an explosion of recent work on evolutionary game theory.[3] We see here an approach to the behavior of players in games familiar from the beginnings of economics. The inhabitants of economic models are often represented as maximizers, but this maximization is typically interpreted not as springing to life as the result of sophisticated calculation, but rather as the outcome of a process of trial-and-error adjustment. This type of adjustment takes center stage in evolutionary game theory.

Second, we must now recognize that the models with which we work are not literal descriptions of the strategic interactions of interest, but are indeed *models* of those interactions. We hope that these models capture the essential elements of the interaction and ignore the trivial, but must ever be open to the possibility that something important is missing from the model. If game theory takes us to seemingly counterintuitive conclusions, the difficulty is likely to be found not in the structure of game theory itself but in the models with which it has been implemented.

Third, the choice of an equilibrium is now part of the construction of a model rather than an implication of the model. An equilibrium that is appropriate for some applications of the model may be irrelevant in others.

---

[3] For a taste of this literature, see Fudenberg and Levine (1998), Hofbauer and Sigmund (1988), Mailath (1998), Samuelson (1997), van Damme (1991, ch. 9), Vega Redondo (1996), Weibull (1995) and Young (1998).

## 15.2.3 A Comparison

It will be useful to compare these two points of view in the context of the following game:

$$
\begin{array}{c c}
 & 2 \\
 & \begin{array}{cc} A & B \end{array} \\
1 \begin{array}{c} A \\ B \end{array} & \begin{array}{|c|c|} \hline 1,1 & 0,0 \\ \hline 0,0 & 1,1 \\ \hline \end{array}
\end{array}
$$

This is a coordination game, with three equilibria—one in which both agents choose $A$, one in which they both choose $B$, and one in which they each mix equally over these two alternatives. For some, this multiplicity is a potentially fatal flaw of game theory. How are we to use a model if it leads to multiple outcomes? What empirical content does game theory have if we cannot make its implications precise? This criticism recurs in full force when dealing with repeated games, where the Folk Theorem tells us that virtually anything can be an equilibrium outcome (cf. Mailath and Samuelson [32]).

In the classical view, all of the information associated with this game is contained in the payoff matrix, and nothing beyond this can be used to identify equilibrium behavior. Finding nothing to break the symmetry of the game, Harsanyi and Selten [22] accordingly select the mixed equilibrium as the outcome. One might be tempted to recoil from the inefficiency of this equilibrium, but if one has no history to guide behavior, no labels to distinguish strategies, no shared understanding with one's opponent, no choice can be more effective than mixing.

In the humanistic view, this game comes in a context. Perhaps this game is offered in the course of an analysis of driving habits and the strategies $A$ and $B$ are to be interpreted as the alternatives of driving on the right side or on the left side of the road. For most such applications, one of the two pure equilibria is relevant, while the mixed equilibrium is absurd. Which of the pure equilibria is relevant depends on the circumstances. We observe people driving on the right in some countries and on the left in others. The players coordinate on such an equilibrium, despite the absence of any asymmetries in the model, because they share a history of behavior in similar situations. Which equilibrium emerges in any particular circumstance depends upon accidents of this history that have been excluded from the model. One possibility, of course, would be to incorporate all of this history directly within the model. Unfortunately, this is a certain prescription for models so unwieldy as to be useless. The more productive alternative is to effectively embed this history in the model in the form of an equilibrium selection. However, we must then remember that the job of constructing the model is not finished with the specification of players, strategies and payoffs, requiring also a choice of equilibrium. Laments that games have too many equilibria are then simply reminders that the world is indeed often a complicated place and that we cannot expect an incomplete model to magically do a job that rightly falls to the modeler.

There is a place for both versions of game theory, and there is certainly much to be learned from the pursuit of game theory as a study of rationality. If nothing else, a model of perfectly rational players often serves as a useful point of departure before plunging into the world of human idiosyncracies, much as competitive equilibrium is a useful beginning when studying more complicated markets. At the same time, much of the promise of game theory in economics comes from using game theory to investigate the behavior of people, and I hereafter focus on this humanistic approach.

## 15.3 What is Game Theory About?

What does game theory have to say about such people? Again, it is helpful to distinguish four (interrelated) possibilities.[4]

### 15.3.1 Normative Implications

Game theory can be useful in making prescriptions for how people *should* behave. The quickest route to an idea of how game theory is doing in this regard is to ask whether people value the advice of game theorists. The answer is clearly yes. The most visible evidence is the recent role of game theorists in designing and running auctions for spectrum rights (see Binmore and Klemperer [6], Klemperer [26] and Milgrom [36]), though one also finds the influence of game theory in the allocation of such disparate objects as places in schools (Abdulkadiroğlu, Pathak, Roth and Sönmez [1]), access to railroad tracks (Brewer and Plott [7]), payload priority on the space shuttle (Ledyard, Porter and Wessen [30]), air-

---

[4]Rubinstein [44, pp. 190-191] offers a similar categorization of the uses of economic theory.

port take-off and landing slots (Rassenti, Smith and Bulfin [40]), medical residents (Roth and Peranson [41]), and kidneys (Roth, Sönmez and Ünver [42]).

Game theory has been successful in these applications despite the fact that the settings are typically too complicated to capture within a single model. For example, the simple models that provide our standard results on auction theory are no match for the complexities of the auctions to which they have been applied. However, these models have been useful in identifying principles that can guide our practice in more complicated settings. Indeed, in many cases, game theory has been useful precisely because it makes specific considerations clear in a simple setting—a practical reminder that the most useful model is not always the most detailed.

### 15.3.2 Positive Implications

Game theory can be useful in the analysis of behavior. Hammerstein and Riechert [19] showed that evolutionarily stable strategies are helpful in analyzing the behavior of spiders. Aumann and Maschler [5] explain a series of examples from the Babylonian Talmud in terms of the nucleolus. Young and Burke [52] use evolutionary game theory to analyze the pattern of share-cropping contracts among Illinois farmers. Athey and Haile [4], Haile [16] and Haile and Tamer [17] use game theory to study auctions.

### 15.3.3 Analytical Implications

Game theory allows us to make precise, and hence evaluate and revise, our intuition about economic behavior. A string of such examples early in the 1980s gave rise to the strategic revolution of that decade. For example, economists had long talked of limit pricing—the idea that the prices of an incumbent firm with market power might be set with an eye toward affecting the decisions of potential entrants (cf. Kamien and Schwartz [25]). But why should the prices that a firm sets before rivals have entered the market have any effect on what happens after entry (Friedman [11])? Milgrom and Roberts' [37] demonstration that such a link could arise out of incomplete information, and Harrington's [21] and Mathews and Mirman's [33] exploration of this link, opened the door to the study of such behavior. Similarly, Rubinstein's [43] game-theoretic formulation, making precise the advantage conferring by the ability to impose costly delays on a reluctant opponent, gave birth to the study of (noncooperative) bargaining. Kreps, Milgrom, Roberts and Wilson's [28, 29, 38] study of predatory responses to potential competitors opened the way to the study of reputations. One could cite many more such examples. Each of these models can be criticized for ignoring a wide variety of important features. None is ready for immediate use in either a normative or positive context. But each played a key role in sorting out which of our ideas make sense and which do not. Without the ability to do so, we cannot be assured of the ability to make progress, the ability to dismiss unworkable ideas and intuition in favor of more useful ones, that is characteristic of scientific inquiry.

### 15.3.4 *Conceptual Implications*

Game theory allows us to identify a few key ideas that recur as fundamental in a wide variety of strategic interactions. The prisoners' dilemma has secured its place as the most-studied of games because it captures a tension between individual and collective interests that appears repeatedly in our social lives. Virtually none of the situations said to be prisoners' dilemmas literally match the model – one could not hope to compress a decades-long strategic arms race into a single pair of binary choices – but it is immediately informative to note that the arms race "is" a prisoners' dilemma.

Much the same purpose is served by the other workhorses of game theory, such as the battle-of-the-sexes, matching pennies, coordination games, the hawk-dove game, or the war of attrition. Much of what game theory does is allow us to isolate features that we think are common, though not exclusively descriptive, to different problems. Notice that the very usefulness of these models stems from their starkness. The prisoners' dilemma is so simple that we can find its elements in a vast variety of settings. A more complicated model would be a better fit for any particular setting, but would compromise the ability to find common elements across settings.

## 15.4 Missing Pieces

Much lies ahead for the study of game theory. It is useful to organize comments here around the various applications of game theory identified in the previous section. I will have nothing to say about normative applications, to a large extent because the market appears to be firmly convinced that game theory is a winner in this sense. There is work to be done in each of the other three areas, especially in establishing links between them.

### 15.4.1  Positive Applications

Three points clamor for attention as we develop positive applications of game theory. We need to know more about how we should balance additional complexity against enhanced applicability when constructing models. We need to know more about how we usefully compare theoretical models and observed behavior. We need to know more about how we should select equilibria in the course of modeling a strategic interaction.

First, the overarching principle in thinking about positive applications of game theory is that the models with which we work are by design approximations of the interactions we seek to study. Without doing anything further, we thus know that the models are incorrect,[5] meaning that we cannot expect them to match all of the behavior we observe all of the time. There is then a sense in which those who simply call our attention to situations in which our models perform poorly tell us nothing new. They *do* bring something new to the discussion if they can point the way to enhanced models that will capture more of what we observe. The difficulty here is that such enhanced models are invariably more complex, while models are valuable for their simplicity as well as their applicability. Unfortunately, we have no systematic way to evaluate whether the extra complexity of a model is worth its enhanced ability to match behavior.[6] There is no question that we should be ever on the alert for better models, but also no question that simply adding features to a model and then noting that it better matches the behavior we observe does not by itself make progress.

Second, the fact that our models are approximations leads to some familiar tension in positive applications of the theory. Our

---

[5] Since coming to the attention of economists with their paper on prospect theory [24], Kahneman and Tversky built a research program around showing that economic and game-theoretic models are not always correct, while inspiring others to do likewise. For a striking recent example, see Ariely, Loewenstein, and Prelec [3].

[6] See Harless and Camerer [20] for one such study along these lines.

theoretical work, based on the models, often leads to relatively precise results. The behavior that emerges from the noisy world, incorporating the many complications excluded from the model in the interest of tractability, never matches the model exactly. How are we to put the two together?

The same difficulty appears throughout empirical work in economics. It is standard, for example, to work with an exact theoretical model of a consumer, but then to append an error term when turning to the data. Much the same can be done with game theory, but the self-referential nature of the decisions involved causes the details to become more intricate. Should one simply perturb the behavior that emerges from the model? The techniques for doing so are straightforward, but the results difficult to interpret in terms of the model. Should one instead perturb the model, perhaps perturbing the payoffs of the agents, and then let the model generate the predicted behavior? The difficulty here is that the method for doing so can become quite tightly tied to details of the problem at hand (e.g., Andreoni and Samuelson [2]). Should we perturb behavior at the individual level, and then work through the implications of the theory? The quantal response model of McKelvey and Palfrey [34, 35] provides one quite general way of doing the latter, but without some additional structure, this generality is purchased at the cost of a model that rationalizes any possible behavior (see Haile, Hortacsu and Kosenok [15], and Goeree, Holt and Palfrey [14] for a modification of quantal response equilibrium that imposes some such structure). In general, how are we to assess when observed behavior is sufficiently consistent with a model to reinforce our assessment that the model is useful, and when should it push us to reconsider?

Third, Section 15.2.2 noted that under a positive view of game theory, selecting an equilibrium is an integral part of constructing one's model. The difficulty here is that we have few principles to guide the selection of equilibrium or to interpret the selection we have made. Nowhere is this tension more apparent than when working with repeated games, though it regularly arises elsewhere as well. For example, it is common to restrict attention to Markov equilibria when examining dynamic games – presumably to simplify the analysis – but not when doing repeated games – otherwise we lose interesting implications from the repetition – even though the latter are typically interpreted as convenient approximations of the former. It is common to suggest that attention should be restricted to simple strategies, unless efficiency, another commonly-invoked criterion, requires complicated strategies. To some extent, this is a reflection of game theory's history. We have years of experience in thinking about choosing a game so as to capture the features of interest, but are less accustomed to thinking of equilibria in the same way. Our positive uses of game theory will become more effective as we become more accustomed to this view of equilibria.

## 15.4.2 Analytical Applications

Game theory provides a way to make precise our intuition about how people interact, allowing us to discard or modify those intuitions that do not withstand a rigorous representation. However, intuition is not always easily captured, sometimes because it is faulty, other times because it is not precisely formulated, and still other times because it is more subtle than first appreciated. There are still many ideas that lie at the heart of modern game theory and that appeal compellingly to our intuition but which we do not understand, and that call for better models.

First, we now have a rich theory of repeated games (see Mailath and Samuelson [32] for an introduction). This has allowed us to explore in some detail the idea that links between current and future actions can create current incentives that otherwise could not arise, with profound effects for equilibrium behavior and payoffs. At the same time, it seems unlikely that players ever play literally the same game over and over. Instead, we typically view a repeated game as an approximation of a situation in which people repeatedly play ever-changing but similar games, i.e., in which people play a dynamic game. How faithful is the work on repeated games to this more challenging setting? We have relatively complete results for dynamic games of perfect information, but know much less about dynamic games of imperfect monitoring.

Second, one encounters the idea of a reputation throughout economics. Firms are said to have reputations for providing good service, politicians for being free from corruption, news sources for being unbiased, individuals for being honest, and so on. The key to such reputations appears to be a link between current behavior and expectations of future behavior. Game theorists have thus naturally turned to repeated games as the natural platform for modelling reputations, centering the analysis around repeated games of incomplete information. The resulting adverse selection necessarily introduces a link between current behavior and future expectations, but is only one means to such an end rather than being obviously the most effective way to capture reputations. There remains much that we do not understand about what reputations are and how they work, and hence much yet to be done (see, for example, Mailath and Samuelson [32, Chapter 18]).

Third, Schelling [46] focussed attention on the idea of a commitment, showing with a number of deftly-chosen examples that the ability to make commitments can be valuable. Game theorists typically model commitments as an ability to observably move first. A firm is said to be able to make a commitment to its production plan if it makes its choice first and if its opponents can see this choice. While this often yields tractable models, it places an uncomfortable stress on details of timing that one often suspects are insignificant. People understand what it means when told that someone is committed to a course of action, without requiring that all of the choices involved in that course of action have already been made and documented. We understand the sense in which some politicians are committed to certain ideas while others appear to have no commitment at all. We similarly understand that there are some things to which people can commit, and others to which they cannot, accordingly to patterns that vary with the person and the setting. What lies behind such commitments and how can we use the structure of commitments to more effectively design economic and social institutions? There is clearly some overlap here with work on reputations, and clearly more to be done.

Fourth, concerns about commitment spill over into our models of bargaining. Following Rubinstein [43], the standard noncooperative model of bargaining posits rules for how offers can be made and accepted (or rejected and countered), with an agreement ending the game. Why does an agreement end the game? One possible answer is that the agreement constitutes a contract that can be enforced by an outside agency. This might be appropriate if a firm and union are bargaining or if one is bargaining over the sale of a house. In each such case, the courts stand ready in the event of a dispute. However, bargaining models are also routinely applied to settings in which there is no such authority. For one example among many, economic analyses of families are routinely built around bargaining models. Though courts are often invoked in the event of disputes sufficiently rancorous as to provoke divorce, they are typically not involved in monitoring ongoing relationships. Still, our models are written as if an agreement constitutes a binding commitment. What makes this commitment binding? Again, our analysis has not yet captured all of our intuition.

Fifth, the concept of authority runs throughout discussions of social and political organization, but again eludes our models. It seems apparent that it makes sense to ask who has authority and who does not in any given setting, to view this authority as important, and to think there are things we can do to modify the authority structure in our society. However, we have little idea about what confers authority and why it is valuable. For example, one might view a constitution as a specification of who has authority. However, it is commonly noted that the constitution of Liberia is modeled on that of the United States, though daily life in these two countries is quite different. The constitution of the USSR "guaranteed" a variety of rights, despite a daily life that seemed quite different. In a sense, a constitution is nothing but cheap talk. The wheels of justice can do their job only if the people acquiesce in their doing so. A constitution may focus attention on a desirable equilibrium, but may not. Then what is authority, when it is effective, and how do we make use of it? Is authority simply a way of focussing attention on a particular equilibrium of a game, or is it something more? These are again questions that await models.

## 15.4.3 Conceptual Applications

Game theory has provided us with simple but powerful conceptual models, with perhaps the most notable being the prisoners' dilemma. We often use these models in organizing our thoughts about a strategic interaction, even while recognizing that any such model captures only part of the interaction. An important next step is to recognize that the agents we study may themselves use such models in evaluating their interactions. Just as an analyst relies on simple models to organize her thoughts, so may the agents to whom these models are meant to apply. Moreover, the analyst and the agent need not always think in terms of the same model.

For example, an analyst who offers people choices between various pairs of alternatives, each offering a sum of money at a certain date, may have designed the interaction to ensure the selected sum is received and may not even consider the possibility of doubts about such receipt. Those making the choices, however, having grown accustomed to being leery of promises that "I'll gladly pay you next Tuesday," may incorporate very real doubts about payment in their reasoning. The analyst may then interpret the observed behavior as reflecting a present bias in preferences, while the agent may be manifesting uncertainty aversion (e.g., Halevy [18]).[7] Strictly speaking, the agent's model of the interaction is incorrect. However, the agent must be incorrect in some aspects of her analysis, if she is to use a model – and who would vouchsafe the agent the use of a device that analysts have found essential? – and may well have found this type of approximation useful. We must then be wary of drawing from seemingly anomalous behavior in such games inferences about subjects' preferences or other characteristics.

By its very nature, game theory emphasizes the ability of people to model and react to others' behavior. It then should be a natural setting to take seriously that people use models in making their decisions. Nonetheless, this is a conceptual innovation that remains largely undeveloped, though one with potentially far-reaching implications, including implications for the positive applications of game theory.

## 15.5 Conclusion

Game theory must be on to something. One might dismiss its continuing study in mathematics as being without practical application and its spread through economics as a fad, but cannot similarly dismiss the fact that it is now also a standard tool in disciplines as varied as anthropology, biology, computer science, philosophy, and political science, while continually making inroads elsewhere. Game theory has taken center stage across a broad swath of academic inquiry because it provides a framework that is both flexible and precise, powerful and subtle. The framework

---

[7] Similarly, an analyst may focus on the one-shot nature of an isolated interaction, modelling it as the ultimatum game, while an agent may be in the habit of treating all bargaining interactions as if there is always a future, leading to behavior quite different than what one would expect of the ultimatum game. It is not obvious that people take this point of view. For example, experimental behavior in one-shot and repeated games is markedly different (e.g., Fehr and Gächter [9] and Gächter and Falk [13], see also Fehr and Henrich [10]). This strongly suggests that subjects *do* pay attention to the relative likelihood of future interactions when selecting their models. However, it does not imply that they always model this likelihood correctly, and in particular that they do not sometimes use models incorporating some possibility of future interaction to examine single-shot games.

by itself, of course, cannot do the work. Any application of game theory requires a careful mix of modelling and analysis, an art that makes game theory a great deal of fun as well as useful, but an art that is sufficiently subtle as to leave us with a great deal yet to discover.

[1] Atila Abdulkadiroğlu, Parag Pathak, Alvin E. Roth, and Tayfun Sönmez. The boston public school match. *American Economic Review*, 95(2): 368–371, 2005.

[2] James Andreoni and Larry Samuelson. Building rational cooperation. *Journal of Economic Theory*, 2005. Forthcoming.

[3] Dan Ariely, George Loewenstein, and Drazen Prelec. 'Coherent arbitrariness': Stable demand curves without stable preferences. *Quarterly Journal of Economics*, 118: 73–106, 2003.

[4] Susan Athey and Phil Haile. Empirical models of auctions. In Whitney K. Newey, Torsten Persson, and Richard Blundell, editors, *Advances in Economics and Econometrics: Theory and Applications, Ninth World Congress*. Cambridge University Press, Cambridge, 2006.

[5] Robert J. Aumann and Michael Maschler. Game theoretic analysis of a bankruptcy problem from the talmud. *Journal of Economic Theory*, 36(2): 195–213, 1985.

[6] Ken Binmore and Paul Klemperer. The biggest auction ever: The sale of the British 3G telecom licences. *Economic Journal*, 112(478): C74–C96, 2002.

[7] Paul J. Brewer and Charles R. Plott. A binay conflict ascending price (BICAP) mechanism for the decentalized allocation of the right to use railroad tracks. *International Journal of Industrial Organization*, 14: 857–886, 1996.

[8] Eddie Dekel and Farul Gul. Rationality and knowledge. In D. M. Kreps and K. F. Wallis, editors, *Advances in Economics and Econometrics: Theory and Applications, Volume I*, pages 87–172. Cambridge University Press, Cambridge, 1997.

[9] Ernst Fehr and Simon Gächter. Cooperation and punishment in public goods experments. *American Economic Review*, 90: 980–994, 2000.

[10] Ernst Fehr and Joseph Henrich. Is strong reciprocity a maladaptation? On the evolutionary foundations of human altruism.

In Peter Hammerstein, editor, *The Genetic and Cultural Evolution of Cooperation*, pages 55–82. MIT Press, Cambridge, Massachusetts, 2003.

[11] James W. Friedman. On entry preventing behavior. In S. J. Brams, A. Schotter, and G. Schwodiauer, editors, *Applied Game Theory*, pages 236–253. Physica–Verlag, Vienna, 1979.

[12] Drew Fudenberg and David K. Levine. *Theory of Learning in Games*. MIT Press, Cambridge, 1998.

[13] Simon Gächter and Armin Falk. Reputation and reciprocity: Consequences for the labour relation. *Scandivanian Journal of Economics*, 104: 1–26, 2002.

[14] Jacob K. Goeree, Charles A. Holt, and Thomas R. Palfrey. Regular quantal response equilibrium. Mimeo, California Institute of Technology and University of Virginia, 1004.

[15] Philip Haile, Ali Hortacsu, and Grigory Kosenok. On the empirical content of quantal response equilibrium. Mimeo, Yale University, 2004.

[16] Philip A. Haile. Auctions with resale markets: An application to U. S. forest service timber sales. *American Economic Review*, 92(3): 399–429, 2001.

[17] Philip A. Haile and Elie Tamer. Inference with an incomplete model of english auctions. *Journal of Political Economy*, 111(1): 1–51, 2003.

[18] Yoram Halevy. *Strotz meets Allais: Diminshing impatience and the certainty effect*. Mimeo, University of British Columbia, 2005.

[19] Peter Hammerstein and Susan E. Riechert. Payoffs and strategies in territorial contests: ESS analyses of two ecotypes of the spider em Agelenopsis aperty. *Evolutionary Ecology*, 2: 115–138, 1988.

[20] David W. Harless and Colin F. Camerer. The predictive utility of generalized expected utility theories. *Econometrica*, 62:1251–1290, 1994.

[21] Joseph E. Harrington, Jr. Limit pricing when the potential entrant is uncertain of its cost function. *Econometrica*, 54: 429–437, 1986.

[22] John C. Harsanyi and Reinhard Selten. *A General Theory of Equilibrium Selection in Games*. MIT Press, Cambridge, Massachusetts, 1988.

[23] J. Hofbauer and K. Sigmund. *Evolutionary Games and Population Dynamics*. Cambridge University Press, Cambridge, 1998.

[24] Daniel Kahneman and Amos Tversky. Prospect theory: An analysis of decision under risk. *Econometrica*, 47: 263–291, 1979.

[25] Morton I. Kamien and Nancy L. Schwartz. Limit pricing and uncertain entry. *Econometrica*, 39(3): 441–454, 1971.

[26] Paul Klemperer. *Auctions: Theory and Practice*. Princeton University Press, Princeton, 2004.

[27] Elon Kohlberg and Jean-Francois Mertens. On the strategic stability of equilibria. *Econometrica*, 54: 1003–1038, 1986.

[28] David M. Kreps, Paul R. Milgrom, John Roberts, and Robert J. Wilson. Rational cooperation in the finitely repeated prisoners' dilemma. *Journal of Economic Theory*, 27(2): 245–252, 1982.

[29] David M. Kreps and Robert J. Wilson. Reputation and imperfect information. *Journal of Economic Theory*, 27(2): 253–279, 1982.

[30] John O. Ledyard, David Porter, and Randii Wessen. A market-based mechanism for allocating space shuttle secondary payload priority. *Experimental Economics*, 2: 173–195, 2002.

[31] George J. Mailath. Do people play Nash equilibrium? Lessons from evolutionary game theory. *Journal of Economic Literature*, 36: 1347–1374, 1998.

[32] George J. Mailath and Larry Samuelson. *Repeated Games and Reputations: Long-Run Relationships*. Oxford University Press, Oxford, 2006.

[33] Steven A. Matthews and Leonard J. Mirman. Equilibrium limit pricing: The effects of private information and stochastic demand. *Econometrica*, 51(4): 981–996, 1983.

[34] Richard D. McKelvey and Thomas R. Palfrey. Quantal response equilibria for normal form games. *Games and Economic Behavior*, 10: 6–39, 1995.

[35] Richard D. McKelvey and Thomas R. Palfrey. Quantal response equilibria in extensive form games. *Experimental Economics*, 1: 9–41, 1998.

[36] Paul R. Milgrom. *Putting Auction Theory to Work*. Cambridge University Press, Cambridge, 2004.

[37] Paul R. Milgrom and John Roberts. Limit pricing and entry under incomplete information: An equilibrium analysis. *Econometrica*, 50: 443–460, 1982.

[38] Paul R. Milgrom and John Roberts. Predation, reputation and entry deterrence. *Journal of Economic Theory*, 27(2): 280–312, 1982.

[39] John F. Nash. Non-cooperative games. *Annals of Mathematics*, 54(1): 286–295, 1951.

[40] S. J. Rassenti, V. L. Smith, and R. L. Bulfin. A combinatorial auction mechanism for airport time slot allocation. *Bell Journal of Economics*, 13: 402–417, 1982.

[41] Alvin E. Roth and E. Peranson. The redesign of the matching market for american physicians: Some engineering aspects of economic design. *American Economic Review*, 89: 748–780, 1999.

[42] Alvin E. Roth, Tayfun Sönmez, and Utku Ünver. A kidney exchange clearinghouse in new england. *American Economic Review*, 95, 2005.

[43] Ariel Rubinstein. Perfect equilibrium in a bargaining model. *Econometrica*, 50(1): 97–109, 1982.

[44] Ariel Rubinstein. *Modeling Bounded Rationality*. MIT Press, Cambridge, 1998.

[45] Larry Samuelson. *Evolutionary Games and Equilibrium Selection*. MIT Press, Cambridge, 1997.

[46] Thomas Schelling. *The Strategy of Conflict*. Harvard University Press, Cambridge, MA., 1980 (first edition 1960).

[47] Eric van Damme. *Stability and Perfection of Nash Equilibria*. Springer-Verlag, Berlin, 1991.

[48] Fernando Vega-Redondo. *Evolution, Games, and Economic Behavior*. Oxford University Press, Oxford, 1996.

[49] John von Neumann and Oskar Morgenstern. *Theory of Games and Economic Behavior*. Princeton University Press, Princeton, 1944.

[50] Jörgen W. Weibull. *Evolutionary Game Theory*. MIT Press, Cambridge, 1995.

[51] Peyton Young. *Individual Strategy and Social Structure*. Princeton University Press, Princeton, 1998.

[52] Peyton Young and Mary A. Burke. Competition and custom in economic contracts: A case study of illinois agriculture. *American Economic Review*, 91: 559–573, 2001.

# 16
# Thomas C. Schelling

Lucius N. Littaluer Professor of Political Economy, Emeritus
Harvard University, USA
Distingished University Professor
University of Maryland, USA

---

Why were you initially drawn to game theory?

In the late 1940s and early 1950s I was a participant in international negotiations and developed some ideas that, when I left Government in 1953 to join the faculty of Yale, I thought I'd work on. Eventually I published an "Essay on Bargaining" in which I explored the concept of "commitment," how one may (or may fail to) adopt a bargaining position that he or she is obliged or motivated to adhere to and – crucially important – communicate that obligation or motivation in a credible manner to another party. I explored promises, threats, and bargaining tactics, looked at contracts, reputation, appeals to a deity, physical positioning (burning bridges), uses of agents.

I also looked at coordination when payoffs to two or more parties were identical but equilibria were multiple (or infinite), and when payoffs were asymmetrical but dependent on coordination. I was convinced that coordination of expectations was often crucial to the completion of overt negotiation.

I had just enough acquaintance with game theory to realize that what I was doing might be construed as game theory, but not enough to be tempted to formalize any of my ideas as game theory. In late 1957, after I had finished that work, I came upon *Games and Decisions* (1957), by R. Duncan Luce and Howard Raiffa, and spent a hundred hours learning game theory.

Immediately after, I spent eight months in London on a fellowship and worked further on the same kinds of ideas, with some intention to relate them explicitly to game theory, having absorbed enough from the Luce-Raiffa book to feel sure of the connection. I felt that game theory, as I had come across it, was more abstract than it needed to be and could usefully be enlarged in scope to encompass strategies and tactics of bargaining, becoming empirical and historical as well as logical and mathematical. I was presumptuous enough to subtitle my article, as published in the *Journal of Conflict Resolution* (1958), "Prospectus for a Reorientation of Game Theory." (Of course, nobody took it up.) In that article I used a few payoff matrices, mostly 2 × 2, without thinking that that was what made it game theory.

It was during my stay in London that I worked on a problem that intrigued me, which I called the "reciprocal fear of surprise attack," and began to realize that the reciprocal deterrence between the USA and the USSR was the highest priority for my interests. I had agreed to join the RAND Corporation for a year, and during 1958–59 I began a professional interest in nuclear-weapons policy and arms control that preoccupied me for the next decade. I did, at RAND, produce a couple of articles (included in my 1960 book, *The Strategy of Conflict*) that were explicitly game theory. Mainly I learned about nuclear weapons technology and policy.

I then produced many articles and two books on nuclear defense strategy and arms control. None of that work looked like "game theory," although it was of the same nature as my earlier work that used some game-theoretical terminology and simple matrices. I also worked on criminal coercion and "organized crime," on somewhat game-theoretic strategies of self-management or self control, on racial segregation, and on multi-person interactions like self-confirming expectations, multi-person prisoners' dilemmas, dual-equilibrium binary multi-person choices, and situations embodying the "fallacy of composition." I think the latter work can be construed as many-person game theory but in no way depended on formal game theory.

I was surprised and somewhat perplexed when the Nobel selection committee for economic sciences awarded me the Bank of Sweden Prize in Memory of Alfred Nobel in 2005 "for having deepened our understanding of conflict and cooperation through game-theoretic analysis." I thought I had contributed to an understanding of conflict and cooperation; I thought in 1958 my work might usefully be construed as game-theoretic analysis; I thought in 2005 that, excepting possibly my development of coordination theory and "focal points," what I did was not recognizable as game theory. Maybe the committee was trying to redefine game theory by incorporating my work, much as I had futilely tried in 1958.

Still, if game theory is to be identified with the formal logic of rationally identifying and choosing strategies in equilibrium, and its attendant definition of payoffs and use of matrix notation, I am a user of game theory, not a creator. In most disciplines there is a distinction between "theorists" and professionals, or practitioners. There are economists but also economic theorists; there are sociologists but also sociological theorists; there are statisticians and statistical theorists, even physicists and theoretical physicists. But game theory, unlike economics, or sociology, or statistics or physics, has "theory" in its name. We don't have a term like "gameist" for the one who uses game theory, the way economists use economic theory without necessarily producing theory as economic "theorists" do.

I believe I can distinguish what I do from what game theorists do in the following two ways. One is that most game theory is concerned with identifying rational choice when the optimal choice depends on the choice, or choices, that another is, or others are, anticipated to make. Except for my work on coordination theory, I have been, I believe, almost entirely concerned with how individuals rationally attempt to influence, not to anticipate, the choices of others. And, second, while I have tried to identify the logic of tactics of influence—unilateral promises, reciprocal promises, threats, commitments, the elimination of options, hostages, contracts, appeals to higher authority, etc. I have been mainly concerned with empirical (or sometimes fictional) and historical evidence of behavior. I have been more "descriptive" than "normative".

Most game theory considers such things as commitments, promises, threats, contracts, etc., to be either enforceable or not enforceable; I have been mainly concerned with how and where and by whom in what institutional environments threats, promises, and commitments can be successfully incurred, or successfully bluffed, or successfully countered. I am more social scientist than logician.

## What is the proper role of game theory in relation to other disciplines?

Game theory is like logic or mathematics. It should be available to inform economics, sociology, social psychology, law, anthropology. As such it should be accepted as potentially descriptive; should be adapted to the relaxation of strict requirements of rationality.

The award of the Nobel Prize in economics to Daniel Kahneman suggests how game theory can incorporate logical and other errors in decision to make it more descriptively helpful to the social sciences (including law).

## What do you consider the most neglected topics and/or contributions in late 20th century game theory?

There is a huge difference, I believe, between what game theorists do and what social scientists do with the help of game theory. Social scientists have used game-theoretic frameworks for laboratory studies of interaction; social scientists and historians have used game-theoretic ideas for strategic, political, and historical analysis; legal and legislative scholars have found game theory helpful in identifying principles of law and regulation; military strategists have found game theory – and here I do not refer to zero-sum theory, which had early, limited application – helpful in analyzing conflict, cooperation, and alliance. I believe the most fruitful applications of game theory have been by social scientists who do not consider themselves game theorists. Maybe there is a useful division of labor here: game theorists do what they do, and others make use of what they do. It is my observation that the most useful applications of game theory by people who are not fully identified with game theory are by economists, who are probably more intellectually congenial with game theory methodology than most social scientists. If there are "neglected topics" it may not be game theorists who are doing the neglecting.

## What are the most important open problems in game theory and what are the prospects for progress?

Probably the most important problems to which game theory might make progress are not problems in game theory but problems in the social sciences to which rudimentary game theory can provide intellectual guidance and stimulus.

# 17
# Brian Skyrms

UCI Distinguished Professor of Logic
and Philosophy of Science and Economics
Director: Interdisciplinary Program in History and Philosophy of Science
University of California Irvine, USA

---

### Why were you initially drawn to game theory?

Because it makes rational deliberation more interesting when deliberators interact.

### What example(s) from your work (or the work of others) illustrates the use of game theory for foundational studies and/or applications?

Modern Game Theory has really begun to illuminate the dynamics of social norms and conventions – indeed all sorts of problems traditionally addressed in philosophy by social contract theory. I would like to mention Ken Binmore's two volumes on *Game Theory and the Social Contract*, Peyton Young's *Individual Strategy and Social Structure* and, Cristina Bicchieri's *Rationality and Coordination* and *The Grammar of Society*. There are many others. I am leaving out. Among my books there are *Evolution of the Social Contract* and *The Stag Hunt*.

### What is the proper role of game theory in relation to other disciplines?

In evolutionary game theory it is the mathematical theory of evolution with frequency dependant fitness. As such it has applications everywhere from sex ratios to evolution of signals. In the social sciences it is a theory of interactive decisions—rational and irrational. It has applications throughout Economics, Political Science, Sociology and Anthropology.

## What do you consider the most neglected topics and/or contributions in late 20th century game theory?

Games played on network structures, and the dynamics of network formation and evolution, are topics that were largely neglected in the 20th century. Now they are rapidly developing areas.

## What are the most important open problems in game theory and what are the prospects for progress?

The further development of learning dynamics in games could be further integrated with psychology, and more sophisticated cognitive models of learning involving analogy and generalization could be explored.

# 18
# Robert Sugden

Professor of Economics
University of East Anglia, Norwich, UK

Why were you initially drawn to game theory?

Let me begin by saying that I have never thought of myself as a game theorist. When I have to produce a short biography or curriculum vitae, I usually describe my research interests as some mix of economic theory, normative economics, behavioural economics, experimental economics, and philosophy of economics – but not game theory. Certainly, in just about all the areas of economics in which I do claim an interest, my work uses some of the conceptual framework of game theory; but the uses to which I put this framework will often seem idiosyncratic to professional game theorists. That is because I am building theories which are intended to throw light on particular areas of economic and social experience; in theory-building, I use whatever components I find useful, and adapt them to the job in hand. Only rarely do I see myself as contributing to game theory in the generic sense. So, in answering your five questions, I cannot pretend to the status of one of the Great and the Good of game theory. My judgements about game theory as a whole are those of an outsider.

As far as I can recall, game-theoretic themes began to enter my work in the mid-1970s, when my main research interests were in welfare economics and social choice theory. Like many social choice theorists of this era, I was intrigued by Amartya Sen's theorem of 'the impossibility of a Paretian liberal'.[1] 'Social choice' was generally understood in terms of relationships between the preferences of the various individual members of society and what were called 'social preferences'; social choice theory examined different ways of constructing social preferences out of profiles of individual preferences. Sen's impossibility theorem worked by representing a concern for liberty as a particular relationship between individual and social preferences: roughly, in matters deemed private to particular individuals, social preferences should coincide with the preferences of the relevant individual. Following the conventions of social choice theory, social preferences were required to be consistent with the Pareto principle, and to satisfy certain conditions of internal consistency, analogous with the properties of rational consistency usually attributed to individual preferences.

It seemed to me (and still seems to me) that this way of thinking about liberty is misconceived. The idea of consistent social preferences belongs with a conception of normative discourse in which the world is seen from the viewpoint of an imagined moral observer, and in which recommendations are addressed to an imagined social planner or benevolent despot. In contrast, liberty requires a framework of rules within which individuals act as *they* severally choose, in pursuit of *their* separate ends. In approving this kind of process, a liberal does not thereby judge its outcomes to be good in themselves. For such a liberal, then, there are no 'social preferences' over outcomes which can sensibly be required to be consistent. In struggling to express this critique, I came to see that social choice theory was an adaptation of the neoclassical idea that economics is all about maximisation; the idea of social preference was an adaptation of the idea of a social welfare function, the maximisation of which had been taken to be the subject-matter of welfare economics. Liberty, on the other hand, has to be understood as a property of the rules of a *game* (strictly speaking, a game form) in which the members of society are players.[2] Rationality can properly be attributed to individual game-players (or so I thought at first), but we have no reason to expect the outcomes of strategic interactions between rational individuals to have the consistency properties that one might attribute to a rational social planner.[3]

This work on liberty was noticed by James Buchanan, who had been thinking on similar lines. I visited his research centre at the Virginia Polytechnic Institute and was strongly influenced by his *contractarian* approach to normative questions. The essential idea is to distinguish between day-to-day collective choices, which are the composition of choices made by separate individuals acting within a framework of rules (as in my preferred account of liberty), and constitutional choices, which set the rules themselves. For Buchanan, the normative criterion for constitutional choice is *agreement* among the individuals who will be subject to the rules.[4] In the early 1980s I set myself the task of writing a book which would develop this form of contractarianism in relation to the better-known social contract theories of John Harsanyi and John Rawls.

Given the contractarian criterion of agreement, it seems inescapable that an analysis of constitutional choice must be an analysis of *bargaining*. However, I discovered that Harsanyi's and Rawls's theories were constructed in ways that removed the element of bargaining from the formation of a social contract. Rather like Amataya Sen in his analysis of liberty, Harsanyi and Rawls were analysing the interaction of separate individuals as if the whole process was governed by a single mind. This was made possible by two complementary devices. The first was to place the contracting parties behind a 'veil of ignorance' so thick as to prevent them from being aware of any differences between them. The second was to attribute rationality to each of the contracting parties, and to postulate principles of rational choice strong enough to produce a unique solution to their (individual and collective) choice problem. Thus, the collective problem of *agreeing* on welfare judgments (for Harsanyi) or constitutional principles (for Rawls) was reduced to a problem of individual choice under uncertainty.[5]

Neither of these devices seemed suitable for my purposes. In the spirit of Buchanan's work, I wanted to be able to say something about the *actual* acceptability of constitutional rules in *actual* societies. Given this aim, I needed to analyse the case of individuals seeking to agree on a set of constitutional rules, not knowing exactly how different rules would turn out to impact on them, but aware of some basic differences between individuals' starting positions. So I turned to bargaining theory, which claimed to provide unique 'solutions', grounded on generic principles of rationality, for mathematically well-specified bargaining problems. But I also found Thomas Schelling's analysis in his *Strategy of Conflict*, in which the outcome of rational bargaining can be critically dependent on the bargainers' common conceptions of 'prominence' and their common recognition of precedents.[6] Schelling seemed (and still seems) to me to be right. But this thought has radical – not to say, disastrous – implications for social contract theory.

Traditional social contract theory sets out to derive normative principles from the hypothetical deliberations of rational agents, who are somehow screened from knowledge of their day-to-day life in an actual society. It is because of this screening that the results of their deliberations can be presented as reflecting an 'unbiased' or 'neutral' point of view, and hence as having normative force. But if Schelling is right, the same screening will remove many of the points of reference that rational individuals use to resolve real bargaining problems. For the problems faced by the ideal agents of social contract theory, there may not be unique rational solutions. But if we remove enough of the screening to allow the contracting parties to find a solution, that solution is likely to inherit the precedents of the real society from which the parties are supposed to have been abstracted; and so the claim to neutrality may be lost.[7] As I came to appreciate the nature of this problem, I began to realise that my project would involve something more than the application of a body of existing game theory. The turning point came when Buchanan suggested that I read David Hume's *Treatise of Human Nature*.[8] In Hume's work, I found a sketch of a form of game theory that was very different from the modern orthodoxy, but suited to my purposes. I even found answers to some of the specific problems I had been struggling with. To the extent that I am a game theorist at all, that is how I became one.

## What example(s) from your work (or the work of others) illustrates the use of game theory for foundational studies and/or applications?

At the risk of self-indulgence, I use the continuation of this story as the first part of my answer. The book that I had originally planned as a contribution to social contract theory mutated into *The Economics of Rights, Co-operation and Welfare*, an analysis of the emergence and stability of conventions and norms.[9] I like to think that this book (which, for short, I will call *ERCW*) illustrates how game theory can illuminate foundational issues in economics, social theory and moral philosophy. As a further illustration, I will discuss a much more recent book, not written by me, which deals with similar topics. These illustrations introduce themes that are relevant for my answers to the remaining questions.

Re-reading *ERCW* in 2003 when preparing the second edition, I was struck by how little use I had made of orthodox game theory. In the 1980s, the amount of game theory that general economists were expected to know was quite limited, and my degree of knowledge was probably typical. My work drew on theoretical ideas from an eclectic range of sources. I have already mentioned Hume's *Treatise of Human Nature*, which presents the outlines of an analysis of how conventions of coordination and reciprocal cooperation can emerge from the repeated interactions of individuals, each pursuing his own interests. Hume argues that these conventions are, to some degree at least, contingent on particularities of the societies in which they evolve, as viewed through the lenses of human 'imagination'; as a result, they cannot be fully explained as the outcomes of abstract rational deliberation. Hume goes on to argue that our ideas of justice have grown up around such conventions; thus, those ideas too cannot be reconstructed by abstract rational analysis. Using Hume's ideas in conjunction with those of Adam Smith's *Theory of Moral Sentiments*, *ERCW* develops a naturalistic account of morality as a spontaneous order of convention-based sentiments.[10]

*ERCW* integrates these eighteenth-century ideas with game-theoretic analysis from three more modern sources, none of which was altogether orthodox in 1986. The first of these was Schelling's analysis, whose significance for my project I have already explained. The second was a work of philosophy: David Lewis's *Convention*.[11] This book already had classic status among philosophers, and was known by repute to game theorists for its pioneering analysis of common knowledge. But the depth, originality and sophistication of Lewis's game theory was not, and still is not, properly appreciated by game theorists. Among the distinctive features of Lewis's game theory are: its focus on games that are played recurrently within populations, rather than the one-shot games that most game theorists studied in the 1970s and 1980s; its attempt to integrate Schelling's findings about focal points into an analysis of rational play; its attempt to make explicit the actual processes of reasoning that players use, rather than focusing on supposed equilibrium properties of rational play; and (a particular instance of the previous feature) its analysis of *how*, within a population, common knowledge of a proposition can be generated.[12]

My third source was, at the time, by far the least orthodox: the work of the biologist John Maynard Smith and his collaborators, investigating 'the logic of animal conflicts'.[13] When I was writing *ERCW*, game theory was almost universally understood by economists as the analysis of strategic interactions among *perfectly rational* agents; to the extent that ordinary human decision-makers fell short of perfect rationality, their behaviour was deemed to lie outside the domain of game theory. The idea that game theory might be applied to animal behaviour, and that economics might learn something from these applications, had hardly impinged on the consciousness of economists. *ERCW* was a very early example of what has since become a genre in economics: an attempt to adapt the methods of evolutionary game theory, as originally developed in biology, to explain economic and social phenomena. My aim was not to offer biological explanations of these phenomena. Rather, I used evolutionary game theory as a template for developing simple dynamic models of recurrent interactions in human populations; in place of natural selection, I assumed that my agents learned by experience. A significant feature of many of these models of recurrent games was a tendency for prominent but apparently irrelevant asymmetries between players' roles to 'seed' conventions in which those asymmetries were used as coordinating devices. *ERCW* tries to show how the learning processes through which conventions emerge can be influenced by individuals' perceptions of prominence, understood in Schelling's sense.

My second illustration is the work of an interdisciplinary research collaboration, published as the book *Moral Sentiments and Material Interests* (for short, *MSMI*), edited by Herbert Gintis, Samuel Bowles, Robert Boyd and Ernst Fehr.[14] The editors propose the hypothesis that a significant proportion of people are motivated by *strong reciprocity*. A strong reciprocator is motivated to benefit those who benefit him and to harm those who harm him, but has the additional property of being an *altruistic third-party punisher*: he is willing to incur material costs in order to inflict harm on individuals who deviate from social norms, even when those deviations are not specifically directed at him.

This concept of reciprocity was originally developed by Fehr and his collaborators to explain a body of experimental observations. In an experimental design that is now widely used in economics, each member of a small group has the option of making voluntary contributions to a good that is public to the group; the dominant strategy is to contribute nothing, but everyone benefits if everyone contributes. If this is played as a one-shot game, significant contributions are made, but these usually decline steeply if the game is repeated in a succession of rounds. Fehr has developed a variant design in which, after each round, each player can punish any other player at a small but non-zero cost. The existence of this third-party punishment option often induces high and stable levels of contribution. The premise of *MSMI* is that Fehr's experiments are revealing a previously neglected property of human psychology which plays a fundamental role in social organisation. I should say that, while I admire the work reported in *MSMI*, I am not wholly convinced of the truth of this ambitious claim.

For me, the most interesting chapters of *MSMI* investigate whether the existence of a motivation towards strong reciprocity can be explained as the product of biological or cultural evolution. The problem, of course, is that altruistic third-party punishment is *altruistic*: it is costly for the punisher. Could such behaviour persist in a population? *MSMI* addresses this question by investigating game-theoretic models of biological and cultural evolution, looking for mechanisms of group selection which might have operated in hunter-gatherer societies. Some of the authors of *MSMI* argue that cultural group selection is a credible mechanism; others (more bravely) argue that biological group selection may have worked too. Other chapters look at the anthropological evidence. Two authors, Hillard Kaplan and Michael Gurven, present evidence about a prototypical – perhaps *the* prototypical – form of human cooperation: food-sharing in hunter-gatherer societies. The evidence shows that individuals are significant net consumers of food until their mid-teens, with the implication that these societies could not subsist without systematic transfers of food, particularly from adult males to young people and from small families to large. Thus, food-sharing practices cannot be culturally contingent: they must have co-evolved with the distinctively human characteristic of childhood dependency. This conclusion does not imply that hunter-gatherer societies need third-party punishment. (Indeed, Kaplan and Gurven play down this possibility, arguing that food transfers are primarily structured by kinship and direct reciprocity.) But it narrows down a problem that evolutionary game theory needs to solve if it is to explain the emergence of cooperative practices in human societies.

## What is the proper role of game theory in relation to other disciplines?

As will be obvious from the examples I have given, I am convinced that game theory can play an important role in the analysis of problems across the social sciences, and in moral philosophy and philosophy of language (which was the starting-point for Lewis's work). But I have tried to avoid using the word 'application', with its suggestion that game theory is a self-contained body of 'pure' theory, available for other disciplines to *apply*, but not dependent on those other disciplines for corroboration. I believe that one of the biggest obstacles to progress in game theory is the perception or illusion that the theory is concerned with a world of a priori propositions, independent of empirical observation and investigation, but somehow accessible to theorists by intuition.

Of course, there need be no illusion in constructing a self-contained formal analysis of propositions that one takes as intuitive or axiomatic; but I believe it *is* an illusion to imagine that this is a good starting-point for an explanation of strategic interactions between people in the real world. Theorising about actual human behaviour must surely be grounded in, and responsive to, evidence of how human beings think and act. Thus, disciplines which study actual behaviour – for example, economics, psychology, anthropology, linguistics, biology – should not be seen merely as areas in which game theory can be applied. They also provide evidence against which the success of game theory can be judged, and theories of behaviour which game theory might (dare I say it?) *apply*.

One of the clearest signs of game theorists' reluctance to engage with other disciplines on equal terms is the familiar assertion that game theory cannot be brought to bear on a real-world interaction until that interaction has been represented as a formal game, with well-specified payoffs. As outsiders to game theory are often told: Get the payoffs right first. This thought leads easily to another: that game theory cannot be tested by using common-sense interpretations of 'payoff' to represent real-world interactions as formal games, and by then investigating actual behaviour in those interactions. If observed behaviour in the resulting games is found to be contrary to some principle of game-theoretic rationality, game theory reserves the right to deny that the common-sense interpretation of payoff is the correct one. For example, Ken Binmore famously denies that observations of cooperative choices in experimental Prisoner's Dilemma games are evidence of the choice of dominated strategies. He accepts that the material rewards in these experimental games have the structure of the Prisoner's Dilemma payoffs, but argues that the observed behaviour is evidence that the game, when properly described, is not a Prisoner's Dilemma after all.[15]

I accept that game theory does not have to assume that payoffs are the same thing as material rewards. But, I maintain, if it is to be capable of useful application, it must include some theory about *what payoffs are*. That theory should allow us to assign payoffs to the games that we use to model real-world interactions *before* we observe how those games are played. If game theory took this composite form, it would be capable of being tested against evidence. I suspect that many game theorists will agree that 'applications' of the theory require this, but think that the definition of payoffs is a problem for the disciplines that use the theory, not a problem for the theory itself. But it is just this attitude that is the obstacle to progress. Conventional game theory presupposes that each player's motivations can be represented by numerical payoffs, assigned to the outcomes of strategy profiles, and that the combined behaviour of the players of a game can be explained by using solution concepts that use these payoffs as data. But what entitles game theory to claim that this strategy of explanation will work? It is not a self-evident truth that players are motivated by individual payoffs, or that standard solution principles, such as dominance, hold when defined relative to such payoffs.[16] To know if this strategy works, we need to be shown that there *is* a method of assigning payoffs to real-world games such that, when it is used, the solution concepts of game theory lead to successful predictions. If this strategy doesn't work, game theory is at fault and needs to be changed. If progress is to be possible, the first essential is that game theorists recognise that it is *their* job to make their theories fit the world. That is what science is all about.

## What do you consider the most neglected topics and/or contributions in late 20th century game theory?

## What are the most important open problems in game theory and what are the prospects for progress?

What do I consider the most neglected topic in late twentieth-century game theory? In two words: empirical reality. Over the last fifty years, the main strands of work in game theory have been characterised by the systematic avoidance of issues which require empirical investigation.

A particularly striking example, and my nomination for the most unjustly neglected contribution, is Schelling's analysis of focal points.[17] Game theorists have constantly lamented the fact that their solution concepts cannot resolve the problems of 'equilibrium selection' that characterise so many games. They have recognised that, in many real-world games, this problem is solved by the players' use of concepts of prominence, of the kind explained by Schelling. So why has there been so little game-theoretic work on this topic? The answer, I think, is that – as Schelling wrote repeatedly in *Strategy of Conflict* – the analysis of prominence is 'essentially' and 'inherently' empirical.[18] It does not lend itself to the kind of abstract, a priori analysis that can be done without reference to facts about the world. In *Strategy of Conflict*, Schelling presents a sketch of an empirically-based theory of focal points. When I say that this theory has been neglected, I mean more than that game theorists have not developed it: they have been unable to recognise that it is theory at all.[19]

It is strange but true that game theory's preference for the a priori over the empirical has been imported into evolutionary game theory, at least as this is practised by economists. From the late 1980s, there has been a dramatic shift from the idea that game-theoretic equilibrium is a property of ideal rationality and common knowledge to the idea that it is a stationary state of a dynamic process of boundedly-rational learning or cultural transmission. One might have expected that this shift would have been associated with a switch to empirical research methods. An evolutionary social theory needs empirical theories of learning and cultural transmission in the same way that evolutionary biology needs genetics: there must be mechanisms by which variations in behaviour are created, inherited and selected, and the nature of these cannot be discovered by a priori methods. But few evolutionary game theorists have seriously engaged with this kind of enquiry.[20]

It is against this background that *MSMI* stands out as an attempt to develop a form of game theory that is responsive to evidence from experimental economics, psychology and anthropology. Other efforts in this direction include Michael Bacharach's *Beyond Individual Choice* and the body of work reviewed in Colin Camerer's *Behavioral Game Theory*.[21] These may be early signs of a shift in the methodology of game theory, following on from the growth of experimental methods and 'behavioural' approaches to the study of individual choice. I hope so.

## Notes

1. Amartya Sen, 'The impossibility of a Paretian liberal', *Journal of Political Economy* 78 (1970): 152–157.

2. See my 'Social choice and individual liberty', in M.J. Artis and A.R. Nobay (eds), *Contemporary Economic Analysis*, Croom Helm, 1978 and 'Liberty, preference and choice', *Economics and Philosophy* 1 (1985): 213–229.

3. I soon began to have doubts about whether conventional rationality properties should even be attributed to individuals. One of

my first reasons for doubting this was the thought that non-selfish behaviour by individuals might not be the result of altruism (as was generally assumed by economists in the 1970s and 1980s). Instead, it might result from individuals acting as if constrained by moral rules. This thought led me to develop a theory of reciprocity: see my 'Reciprocity: the supply of public goods through voluntary contributions', *Economic Journal*, 94 (1984): 772–787. In parallel, working with Graham Loomes, I began to think about whether the standard axioms of rational choice were as normatively compelling as they were usually thought to be: see, for example, our 'Regret theory: an alternative theory of rational choice under uncertainty', *Economic Journal*, 92 (1982): 805–824.

4. James Buchanan, *The Limits of Liberty*, University of Chicago Press, 1975.

5. John Harsanyi, 'Cardinal utility in welfare economics and in the theory of risk-taking', *Journal of Political Economy* 61 (1953): 434-435; John Rawls, *A Theory of Justice*, Harvard University Press, 1971.

6. Thomas Schelling, *The Strategy of Conflict*, Harvard University Press, 1960.

7. I say more about this problem in my 'Contractarianism and norms', *Ethics* 100 (1990): 768–786.

8. David Hume, *A Treatise of Human Nature*, Clarendon Press, 1978; first published 1739–1740.

9. Robert Sugden, *The Economics of Rights, Cooperation and Welfare*, Basil Blackwell, 1986. Second edition published by Palgrave Macmillan, 2004.

10. Adam Smith, *The Theory of Moral Sentiments*, Clarendon Press, 1976; first published 1759.

11. David Lewis, *Convention: A Philosophical Study*, Harvard University Press, 1969.

12. Robin Cubitt and I explain the significance of Lewis's contribution to game theory in our 'Common knowledge, salience and convention: a reconstruction of David Lewis's game theory', *Economics and Philosophy* 19 (2003): 175–210.

13. John Maynard Smith and G. R. Price, 'The logic of animal conflicts', *Nature* 246 (1973): 15–18; John Maynard Smith, *Evolution and the Theory of Games*, Cambridge University Press, 1982.

14. Herbert Gintis, Samuel Bowles, Robert Boyd and Ernst Fehr (eds), *Moral Sentiments and Material Interests*, MIT Press, 2005.

15. Ken Binmore, *Playing Fair*, MIT Press, 1994, pp. 95–96, 111–113. In my experience, most game theorists endorse Binmore's 'get the payoffs right' principle, even if they disagree with the particular (revealed-preference) definition of payoffs he advocates.

16. As an example of a viable alternative, consider the theory of *team reasoning*, in which individual players are motivated to choose their components of strategy combinations which maximise 'team' payoffs: see Michael Bacharach, *Beyond Individual Choice: Teams and Frames in Game Theory*, Princeton University Press, 2006.

17. Revealingly, focal points were not mentioned at all in the main press release from the Royal Swedish Academy of Sciences when Schelling's Nobel Prize was announced in 2005, and were referred to only briefly in the 'advanced information' provided for specialist readers. The main focus of attention was Schelling's analysis of credible commitments.

18. *Strategy of Conflict*, pp. 97–98, 162–163, 286, 295, 303.

19. For more on the strange neglect of Schelling's theory, see Robert Sugden and Ignacio Zamarrón, 'Finding the key: the riddle of focal points', *Journal of Economic Psychology* 27 (2006): 609–621.

20. For documentation of this critique, see my 'The evolutionary turn in game theory', *Journal of Economic Methodology* 8 (2001): 113–130.

21. Bacharach, *op cit*; Colin Camerer, *Behavioral Game Theory: Experiments in Strategic Interaction*, Princeton University Press, 2003.

# 19
# H. Peyton Young

Scott and Barbara Black Professor of Economics
Johns Hopkins University, USA
Professor of Economics
University of Oxford, UK
Senior Fellow
The Brookings Institution, USA

Why were you initially drawn to game theory?

It would be more accurate to say that I stumbled into game theory because I needed it to solve other problems I was working on. My first brush with the subject occurred in the early 1970s when I was teaching a new course called *Mathematics in the Social Sciences* at the Graduate School of the City University of New York. The topics included voting rules, Arrow's theorem, the Gale-Shapley matching algorithm, the Banzhaf voting power index, the Shapley value, and legislative apportionment, which I was then developing with my colleague Michel Balinski. The course had an applied flavor and involved studying practical methods used for election, apportionment, and college admissions, as well as theories about what methods might be best in principle.

It occurred to me that the *strategic* aspects of voting rules needed to be considered in addition to the usual axiomatic treatments, which had been the dominant paradigm since Arrow's work (Arrow, 1963). This led me to surmise that virtually all voting schemes on three or more alternatives must be manipulable, in the sense that the truthful reporting of preferences may sometimes fail to be a Nash equilibrium. However, I knew virtually nothing about game theory, so I quickly boned up on the subject using Luce and Raiffa's 1957 classic (which is still well worth reading today). I was on my way to proving the result when it occurred to me that someone else might have thought of the idea already. This led me to phone up Kenneth Arrow, who put me on to the papers of Gibbard and Satterthwaite that were just then starting to circulate (Gibbard, 1973; Satterthwaite, 1975).

While this discovery was rather discouraging, and led me to set aside game theory for awhile, it also suggested that it might be profitable to shift attention from negative results in voting theory to positive ones. Impossibility theorems notwithstanding, societies will go on voting by one means or another; the operational question is how to put practical voting methods on a sound theoretical footing. This led me to characterize the classical methods of Borda and Condorcet from first principles, and to show that they can be interpreted as statistical procedures for estimating the choice or ranking that is most likely to be "correct" when voters' judgments are subject to error (Young, 1974, 1975, 1986, 1988).

My next encounter with game theory arose quite by chance in the context of a consulting project. In the early 1980s I was working with two Japanese civil engineers for the municipal water authority of Malmö, Sweden, which wanted to know how to fairly allocate the costs of expansion among its customers. In the course of working on this project, I realized that some of the fairness criteria that I had developed for legislative apportionment could be adapted to this context. A particular example is monotonicity: when the size of the pie to be allocated increases, nobody should get a smaller portion than before. This led to one of the first applied papers in cooperative game theory, showing how ideas like the Shapley value and nucleolus played out in an actual situation with numbers estimated from data (Young, Okada, and Hashimoto, 1982). It also led to my first foray into theory, in which I showed that a simple *marginal contributions principle* could be used to axiomatize both the Shapley value for finite cooperative games and the Aumann-Shapley value for nonatomic games without invoking the additivity axiom (Shapley, 1953; Aumann and Shapley, 1974; Young, 1985a, 1985b).

To sum up, I was initially drawn to game theory not for abstract or philosophical reasons, but because it provided useful tools for solving concrete problems. Moreover, it was *cooperative* game theory and its implications for practical matters of fair division that inspired my earliest theoretical work.

## What example(s) from your work (or the work of others) illustrates the use of game theory for foundational studies and/or applications?

Equilibrium behavior often requires players to use probabilistic strategies in order to create uncertainty in the minds of their opponents. Choice under uncertainty is therefore a central aspect of game theoretic reasoning. But the issue of uncertainty in games goes much deeper than this; indeed, it arises whenever the game has multiple equilibria, for then it is uncertain what equilibrium (if any) will be played. Uncertainty also arises when the players do not know their opponents' utility functions, or even whether their opponents are rational. The issue of what players need to know about a game in order to 'solve' it touches on deep problems in epistemology (Lewis, 1969; Aumann, 1976; Aumann and Brandenburger, 1995). This is one of many instances in which game theory has turned out to have important implications for academic disciplines that at first glance seem far removed from the subject.

In spite of the large demands it places on knowledge and common knowledge, however, many game theorists persist in using Bayesian Nash equilibrium, and refinements thereof, as a central solution concept. I suspect that this is largely because Bayesian rationality offers such a rich playground for clever argumentation. My own feeling is that Bayesian rationality is more of a minefield than a playground. This is not only because it places extraordinary demands on players' reasoning abilities, but also because the conditions under which Bayesian learning actually leads to equilibrium behavior turn out to be very demanding as well.

A number of authors, including Binmore (1987), Jordan (1993), Kalai and Lehrer (1993), and Nachbar (1997, 2005), have examined this issue from different perspectives. Dean Foster and I illustrate the problems with Bayesian learning as follows (Foster and Young, 2001). Suppose that two players are engaged in an infinitely repeated game of matching pennies. Assume that the payoffs are approximately what they would be in a textbook version of matching pennies, but that each of the payoff values is perturbed (once only) by a small random shock. The distribution of the payoff shocks can be common knowledge, but the players do not know their opponents' *realized* payoff values. Suppose that everyone is Bayesian, perfectly rational, and forward-looking. As the game proceeds, can they learn to predict the repeated-game strategy of the opponent, and will play converge to equilibrium play? The answer is: frequently not. It can be shown that, when the shocks are sufficiently small, one or both of the players will *almost surely* fail to learn to predict their opponent's next-period behavior with even approximate accuracy, and period-by-period play will be far from Nash equilibrium play most of the time.

The root of the difficulty is the assumption of exact optimization. An optimizing player who thinks his opponent is playing a mixed strategy can change his behavior very abruptly even though his *beliefs* about the opponent's behavior change only slightly. In a game that has only mixed strategy equilibria, the upshot is that, as each player tries to learn the behavior of his opponent, his own behavior becomes so erratic that *it* is unlearnable. In other words, the interactive nature of the learning process leads to the conclusion that not *both* players can be rational and learn to predict the behavior of the other. This, and related results of Nachbar and Jordan, mean that the use of Bayesian reasoning in games needs to be approached with considerable caution.

## What is the proper role of game theory in relation to other disciplines?

Over the past fifty years game theory invaded first economics, then the rest of social science, and is now colonizing new territory in biology, computer science, and philosophy. One could say without much exaggeration that this is the imperial age of game theory. As in other empires, however, the act of conquest may change the conqueror in unintended ways. Let me venture a couple of predictions along these lines. Up through the 1980s the principal conquests were in economics, specifically the theory of industrial organization and mechanism design. In these settings it seemed reasonable to assume that players are highly rational, interact over long periods of time, and can employ complex forward-looking strategies. This led theorists to focus on the properties of equilibrium in repeated games, which was a major advance over the preceding literature.

I believe that this era is now coming to a close. The fields where game theory is currently having the greatest current impact are computer science, artificial intelligence, and biology, and this will lead to a change of emphasis in the development of theory. In particular, both biology and computer science call for versions of game theory in which 'rationality' plays a less prominent role than in economics. Computer scientists, for example, are concerned with distributed systems of information-processors and how to design protocols that govern their interaction. Game theory is highly relevant to this problem even though the "agents" may not be rational in the sense customarily assumed by economists.

Biology is another subject where game theory is having an important impact, and here too the demands of the application are rebounding on the questions that theorists are asking. The focus of analysis is on the long-run dynamics of large populations of players, whose relative fitness depends on how they are programmed to play. This approach originated in the work of the biologist John Maynard Smith (1982), and has had a substantial impact on the directions that game theory is taking and the ways in which it is being applied in the social sciences.[1] There is no reason why rationality cannot be accommodated in these models; indeed, an interesting open problem is to identify situations where the most highly rational agents in a heterogeneous population actually do have a long-run selective advantage.

My larger point is that game theory's colonization of biology, computer science, and other subjects is having a profound impact on the way that rationality is treated in game theory, on the modeling of out-of-equilibrium dynamics, and a host of other issues in game theory itself.

## What do you consider the most neglected topics and/or contributions in late 20th century game theory?

## What are the most important open problems in game theory and what are the prospects for progress?

I will address these last two questions in tandem. As I mentioned earlier, cooperative game theory is an unjustly neglected topic of research. This was not always the case: von Neumann and Morgenstern put a great deal of emphasis on the cooperative form, and many of the pioneers in game theory made major contributions to the topic (Shapley, 1953; Aumann and Maschler, 1964; Schmeidler, 1969; Aumann and Shapley, 1974). In recent decades, however, the noncooperative approach has increasingly gained the

---

[1] See among others Axelrod, 1984; Foster and Young, 1990; Young, 1993a; Kandori, Mailath, and Rob, 1993; Weibull, 1995; Epstein and Axtell, 1996; Vega-Redondo, 1996; Samuelson, 1997; Fudenberg and Levine, 1998; Young, 1998; Epstein, 2006.

upper hand. Indeed, this trend has gone so far that many textbooks on game theory scarcely give cooperative theory a mention. One reason for this development, as I have already suggested, is that the topics in economics where game theory made its earliest inroads – mechanism design and industrial organization – seem particularly well-suited to the noncooperative approach.

Another reason why cooperative game theory has languished is that its practical applications have not been widely recognized. Earlier I mentioned the problem of sharing costs among the beneficiaries of a public facility. Similar problems arise in setting rates for public utilities (Zajac, 1978). More generally, cooperative game theory is relevant to any situation where scarce resources are to be allocated fairly among a group of claimants. How, for example, should slots at busy airports be allocated among airlines? Which transplant patient should be first in line for the next kidney? How should political representation in a national legislature be fairly divided among parties and geographical regions? Some economists insist that such problems would be solved if they were simply left to the workings of the market. Unfortunately, this overlooks the point that markets are moot unless property rights have been defined and vested in individuals, which is precisely what methods of fair allocation are about.

In my book, *Equity In Theory and Practice* (1994), I examined various fairness concepts from both a foundational and practical standpoint. Cooperative solution concepts like the core and the Shapley value, as well as semi-cooperative notions like the Nash bargaining solution and the Kalai-Smorodinsky solution, provide the entry point for thinking about the meaning of allocative fairness. A close examination of practice, however, suggests that one must go substantially beyond these approaches to formulate a theory that has descriptive validity.

Three central points emerge from the analysis. First, fairness must be judged in the context of the problem at hand. Criteria for allotting transplant organs may be quite different from criteria that pertain to the allocation of legislative seats, and neither may be relevant to the allocation of offices in the workplace or dormitory rooms at college. In other words, notions of justice tend to be compartmentalized and context-specific, a view that has its roots in Aristotelian philosophy, and has been advanced by political philosophers such as Walzer (1983) and Elster (1992).

A second key point is that, in practice, solutions to fairness problems tend to be *decentralized* in the following sense: an allocation is deemed to be fair for a group of claimants only when every subgroup deems that they fairly divide the resources allotted to them. This *subgroup consistency principle* is very ancient. It is implicit, for example, in certain Talmudic doctrines concerning the division of inheritances (Aumann and Maschler, 1985). It also features in many modern solution concepts, such as the core, the nucleolus, and the Nash bargaining solution (Sobolev, 1975; Lensberg, 1988), and in real-world allocation methods such as rules for apportioning seats in legislatures (Balinski and Young, 1982; Young, 1994).

The cooperative game approach to fair division proceeds from an axiomatic standpoint. There is, however, another way of thinking about fairness norms that builds on *noncooperative* game theory. Norms of fair division – indeed norms in general – are often the unpremeditated outcome of historical chance and precedent. What is fair in one society may not be deemed fair in another, because people's expectations are conditioned by precedent, and precedents accumulate through the vagaries of history.

Such processes can be modeled noncooperatively using the framework of evolutionary game theory. As I mentioned earlier, this approach was originally inspired by biological applications, and typically has three key features: i) there is a large population of interacting players; ii) the players have heterogeneous characteristics, including different payoffs, information, and behavioral repertoires, iii) they adapt their behavior based on local conditions and experience, and are purposeful but not always perfectly rational. The focus is on the *dynamics* of such a process, not merely on its equilibrium states. One of the main contributions of the theory is to show that some equilibria have a much higher probability of arising than do others (Foster and Young, 1990; Kandori, Mailath, and Rob, 1993; Young, 1993a). It therefore delivers a theory of equilibrium selection that is based on evolutionary principles rather than on *a priori* principles of 'reasonableness', as in the earlier theory developed by Harsanyi and Selten (1988).

To illustrate how the evolutionary approach can be applied to the study of fairness norms, consider the classical problem of how two individuals would divide a pie. The simplest noncooperative formulation is due to John Nash (1950): each player names a fraction of the pie, and they get their demands provided that both can be satisfied; otherwise they get nothing. Any pair of demands that sums to unity constitutes a noncooperative equilibrium of the one-shot game. If the players are allowed to bargain over time, much tighter predictions are possible. In the standard model, players alternate in making demands, which are either accepted or rejected (Stahl, 1972; Rubinstein, 1982). When the players are perfectly rational and discount future payoffs at the same rate, the outcome of the unique subgame perfect equilibrium is the Nash bargaining solution.

Neither the one-shot demand game nor the alternating offers game is evolutionary in spirit, because they are concerned with what *two particular bargainers* would do in equilibrium, not what a *population of bargainers* would do. To recast the problem in an evolutionary framework, consider a large population of agents who engage in pairwise bargains from time to time. Suppose that the outcomes of previous bargains affect how people bargain in the future, due to the salience of precedent. Once a particular way of dividing the pie becomes entrenched due to custom, people start to think that this is the only fair and proper way to divide the pie, and it therefore continues in force.

To allow for asymmetric interactions, suppose that there are two distinct populations of potential bargainers who are randomly matched each period (e.g., employers and employees). Each matched pair plays the Nash demand game described earlier. Assume for simplicity that all agents in a given population have the same utility function, but that the utility functions differ between populations. To capture the idea that current expectations are shaped by precedent, suppose that each current player looks at a random sample of earlier demands by the opposing side, and chooses a trembled best reply given the sample frequency distribution. (The 'tremble' captures the idea that the process is jostled by small unobserved utility shocks, so that players usually choose a best reply but not always.) It can be shown that, starting from arbitrary initial conditions, players' expectations eventually coalesce around a specific division of the pie, and this endogenously generated norm of division is, with high probability, the Nash bargaining solution. Furthermore, when players are heterogeneous with respect to their degree of risk aversion, a natural generalization of the Nash bargaining solution results (Young, 1993b).

This example shows that there is no need to make extreme assumptions about players' rationality in order for game theory to yield interesting results. Unlike the alternating offers model, where perfect rationality and common knowledge of perfect rationality are assumed, neither is needed in the evolutionary model. Players choose myopic best replies based on fragmentary information, they occasionally make mistakes, and they have no *a priori* knowledge of their opponents' payoffs, behaviors, or degree of rationality. Nevertheless the two models yield essentially the same outcome.

More generally, the evolutionary model of bargaining illustrates how game theory can be used to study the emergence of norms. Over time, interactions among people build up a stock of precedents that may cause their expectations to gravitate toward a particular equilibrium, which then becomes entrenched as a social norm: everyone adheres to it because everyone expects everyone else to adhere to it. When the underlying game is concerned with the division of scarce resources, the resulting equilibrium can be interpreted as a *fairness* norm (Hume, 1739; Binmore, 1994; Young , 1998).

I conclude by hazarding several predictions about the future development of game theory. The first is that rationality, and arguments over how rational the players "really" are, will fade in importance. As I have already argued, game theory can be applied to systems of interacting agents whether or not they are rational in the conventional sense. This insight was initially provided by applications of game theory to biology, and is being buttressed by current applications to computer science, artificial intelligence, and distributed learning.

My second prediction is that game theory will continue to evolve in response to real problems that arise in economics, politics, computing, philosophy, biology and other subjects, a development that von Neumann and Morgenstern would surely have welcomed. While its major successes to date have largely been in economics, game theory is not a sub-discipline of economics; it is more like statistics, a subject in its own right with applications across the academic spectrum.

My third prediction is more of an admonition: game theory will continue to thrive if it remains receptive to new ideas suggested by applications, but risks degenerating if it does not. John von Neumann cautioned about this tendency in mathematics more generally, and game theorists would do well to heed his warning (von Neumann, 1956):

> I think that it is a relatively good approximation to truth – which is much too complicated to allow anything but approximations – that mathematical ideas originate in empirics... As a mathematical discipline travels far from its empirical source, or still more, if it is a second and third generation only indirectly in-

spired by ideas coming from "reality," it is beset with very grave dangers. It becomes more and more purely aestheticizing, more and more purely *l'art pour l'art*. ... [W]henever this stage is reached, the only remedy seems to me to be the rejuvenating return to the source: the reinjection of more or less directly empirical ideas. I am convinced that this was a necessary condition to conserve the freshness and the vitality of the subject and that this will remain equally true in the future.

## References

Arrow, Kenneth J. (1963). *Social Choice and Individual Values*, $2^{nd}$ edition. New Haven, CT: Yale University Press.

Aumann, Robert J. (1976). "Agreeing to disagree," *The Annals of Statistics*, 4: 1236–1239.

Aumann, Robert J., and Adam Brandenburger (1995). "Epistemic conditions for Nash equilibrium," *Econometrica* 63: 1161–1180.

Aumann, Robert J, and Lloyd S. Shapley (1974). *Values of Non-Atomic Games*. Princeton NJ: Princeton University Press.

Aumann, Robert J., and Michael Maschler (1964). " The bargaining set for cooperative games," in Melvin Dresher, Lloyd S. Shapley, and Albert W. Tucker, *Advances in Game Theory, Annals of Mathematics Studies No. 52*. Princeton NJ: Princeton University Press.

Aumann, Robert J. and Michael Maschler (1985). "Game theoretic analysis of a bankruptcy problem from the Talmud," *Journal of Economic Theory*, 36: 195–213.

Axelrod, Robert (1984). *The Evolution of Cooperation*. New York: Basic Books.

Balinski, Michel L., and H. Peyton Young (1982). *Fair Representation: Meeting the Ideal of One Man One Vote*. New Haven: Yale University Press. $2^{nd}$ edition published in 2001 by The Brookings Institution Press, Washington DC.

Binmore, Ken (1987). "Modeling rational players I," *Economics and Philosophy*, 3: 9–55 and "Modeling rational players II", *Economics and Philosophy*, 4: 179–214.

Binmore, Ken (1994). *Game Theory and the Social Contract I: Playing Fair.* Cambridge MA: MIT Press.

Elster, Jon (1992). *Local Justice: How Institutions Allocate Scarce Goods and Necessary Burdens.* New York: Russell Sage Foundation.

Epstein, Joshua (2006). *Generative Social Science.* Princeton NJ: Princeton University Press.

Epstein, Joshua, and Robert Axtell (1996). *Growing Artificial Societies: Social Science from the Bottom Up.* Cambridge MA: MIT Press.

Foster, Dean, and Peyton Young (1990). "Stochastic evolutionary game dynamics," *Theoretical Population Biology*, 38: 219–232.

Foster, Dean P., and H. Peyton Young (2001). "On the impossibility of predicting the behavior of rational players," *Proceedings of the National Academy of Sciences of the USA*, 98: 12848–12853.

Fudenberg, Drew, and David Levine (1998). *Theory of Learning in Games.* Cambridge MA: MIT Press.

Gibbard, Alan (1973). "Manipulation of voting schemes: a general result," *Econometrica* 41: 587–601.

Harsanyi, John, and Reinhard Selten (1988). *A General Theory of Equilibrium Selection in Games.* Cambridge MA: MIT Press.

Hume, David (1739). *A Treatise of Human Nature.* Clarendon Press, Oxford, 1978.

Jordan, James S. (1993). "Three problems in learning mixed-strategy equilibria," *Games and Economic Behavior*, 5: 368–386.

Kalai, Ehud, and Ehud Lehrer (1993). "Rational learning leads to Nash equilibrium," *Econometrica*, 61: 1019–1045.

Kandori, Michihiro, George Mailath, and Rafael Rob (1993). "Learning, mutation, and long-run equilibria in games," *Econometrica*, 61: 29–56.

Lensberg, Terje (1988). "Stability and the Nash solution," *Journal of Economic Theory*, 45: 330–341.

Lewis, David (1969). *Convention: A Philosophical Study.* Cambridge MA: Harvard University Press.

Luce, R. Duncan and Howard Raiffa (1957). *Games and Decisions.* New York: John Wiley.

Maynard Smith, John (1982). *Evolution and the Theory of Games*. Cambridge: Cambridge University Press.

Nachbar, John H. (1997). "Prediction, optimization, and learning in games," *Econometrica*, 65: 275–309.

Nachbar, John H. (2005). "Beliefs in repeated games," *Econometrica*, 73, 459–480.

Nash, John (1950). "The bargaining problem," *Econometrica*, 18: 155–162.

Neumann, John von (1956). "The mathematician," in James R. Newman, *The World of Mathematics*, vol. 4. New York: Simon and Schuster.

Rubinstein, Ariel (1982). "Perfect equilibrium in a bargaining model," *Econometrica*, 50: 97–110.

Samuelson, Larry (1997). *Evolutionary Games and Equilibrium Selection*. Cambridge MA: MIT Press.

Satterthwaite, Mark A. (1975). "Strategy-proofness and Arrow's conditions: existence and correspondence theorems for voting procedures and social welfare functions," *Journal of Economic Theory*, 10: 198–217.

Shapley, Lloyd S. (1953). "A value for n-person games," in Harold W. Kuhn and Albert W. Tucker, eds., *Contributions to the Theory of Games, II, Annals of Mathematics Studies No. 28*. Princeton NJ: Princeton University Press.

Schmeidler, David (1969), "The nucleolus of a characteristic function game," *SIAM Journal on Applied Mathematics*, 17, 1163–70.

Sobolev, A.I. (1975). "Characterization of the principle of optimality for cooperative games through functional equations," in N. N. Vorobyev, ed., *Mathematical Methods in the Social Sciences*. Vilnius: Vipusk 6, 92–151.

Stahl, Ingolf (1972). *Bargaining Theory*. Stockholm: Stockholm School of Economics.

Vega-Redondo, Fernando (1996). *Evolution, Games, and Economic Behaviour*. Oxford UK: Oxford University Press.

Walzer, Michael (1983). *Spheres of Justice*. New York: Basic Books.

Weibull, Jorgen W. (1995). *Evolutionary Game Theory*. Cambridge MA: MIT Press.

Young, H. Peyton (1974). "An axiomatization of Borda's rule," *Journal of Economic Theory*, 9, 43–52.

Young, H. Peyton (1975), "Social choice scoring functions," *SIAM Journal on Applied Mathematics*, 28: 824–838.

Young, H. Peyton (1985a). "Monotonic solutions of cooperative games," *International Journal of Game Theory*, 14: 65–72.

Young, H. Peyton (1985b). "Producer incentives in cost allocation," *Econometrica*, 53: 757–765.

Young, H. Peyton (1986). "Optimal ranking and choice from pairwise comparisons," in Bernard Grofman and Guilllermo Owen, eds., *Information Pooling and Group Decision Making*. Greenwich CT: JAI Press.

Young, H. Peyton (1988). "Condorcet's theory of voting," *American Political Science Review*, 82: 1231–1244.

Young, H. Peyton (1993a). "The evolution of conventions," *Econometrica*, 61: 57–84.

Young, H. Peyton (1993b). "An evolutionary model of bargaining," *Journal of Economic Theory*, 59: 145–168.

Young, H. Peyton (1994). *Equity: In Theory and Practice*. Princeton NJ: Princeton University Press.

Young, H. Peyton (1998). *Individual Strategy and Social Structure*. Princeton NJ: Princeton University Press.

Young, H. Peyton, Norio Okada and T. Hashimoto (1982). "Cost allocation in water resources development," *Water Resources Research*, 18: 361–373.

Zajac, Edward E. (1978). *Fairness or Efficiency: An Introduction to Public Utility Economics*. Cambridge MA: Ballinger.

# About the Editors

**Vincent F. Hendricks** is Professor of Formal Philosophy and member of IIP—Institut Internationale de Philosophie. He is the author of many books including *Mainstream and Formal Epistemology, Thought₂Talk, The Convergence of Scientific Knowledge, Feisty Fragments, Logical Lyrics* and *500 CC: Computer Citations*. Other books include *Self-Reference, Proof Theory, Probability Theory, Interactions, Formal Philosophy, Masses of Formal Philosophy, First-Order Logic Revisited, Philosophy of Mathematics: 5 Questions, Probability Theory and Statistics: 5 Questions* and *Knowledge Contributors*. Editor-in-Chief of *Synthese* and *Synthese Library* he is also the founder of ΦLOG—*The Network for Philosophical Logic and Its Applications* and founding editor of the associated newsletter ΦNEWS.

**Pelle Guldborg Hansen** is doctoral student at the Department of Philosophy and Science Studies at Roskilde University and one of the chief-editors of ΦNEWS.

# About Game Theory: 5 Questions

*Game Theory: 5 Questions* is a collection of short interviews based on 5 questions presented to some of the most influential and prominent scholars in the field. We hear their views on game theory, its aim, scope, use, the future direction of game theory and how their work fits in these respects.

> *This is a terrific collection. The book contains not only interviews with some of the best game theorists of the last decades but also with philosophers and logicians interested in different aspects of the theory of games. The intellectual portraits offered in the collection offer a map of the current status of the discipline and its future directions, including developments in empirical and behavioral economics as well as neuro-economics.*
>
> *This is a must read for anyone interested in contemporary theories of rationality, interactive epistemology and logic.*
>
> — **Horacio Arló-Costa**, Carnegie Mellon University

## WWW.GAMETHEORISTS.COM

© 2007  Automatic Press ♦ VIP

# Index

a priori analysis, 199
Abdulkadiroğlu, A., 170
Abramsky, S., 13
action, 77
adaptionism, 80
adaptive heuristics, 99
adjustment
    trial-and-error, 37, 168
admissibility, 26, 42, 43
    iterated, 43
aggression, 126
agreement, 192
agreement theorem, 3, 99
AIDS, 56
air attack, 1
Alexander, J., 28
algebra, 109
Alice, 129
Allais, M., 149
American Economic Association, 121, 125
analysis, 109
analysis of algorithms, 151
Andreoni, J., 174
anger, 90
animal behavior, 75, 77
animal conflict, 65, 75
animal contest, 78, 83
animal mating, 84
anonymity, 90
anthropologist, 54, 56
anthropology, 58, 125, 178, 187, 190, 198, 200
anthropomorphism, 64
Antonelli, A., 25

appeal to higher authority, 187
apportionment, 204
argument as winning strategy, 10
Aristotle, 10
arithmetic, 25
arms control, 186
arms race, 126, 172
Arrow, K., 122, 204
Arrow's theorem, 203
Arthur Andersen, 113
artificial intelligence, 27, 206
    distributed (DAI), 29
Athey, S.P., 171
Attila the Hun, 38
auction, 6, 50, 87, 119, 144, 170
    design, 35, 119, 170
Aumann, R., 1, 17, 37, 43, 45, 83, 97, 102, 110, 121, 122, 125, 134, 145, 159, 171, 204, 205, 207, 209
    *Handbook of Game Theory*, 3
Aumann, Y., 157
*Aurora*, 147
axiomatics, 109, 209
Ayer, A.J., 62

Babylonian Talmud, 171
Bacharach, M., 53, 200
backward induction, 17, 23
Balinski, M., 203, 209
Banach-Knaster procedure, 146

Bank of Sweden Prize in Memory of Alfred Nobel, 186
Banzhaf voting power index, 203
bargaining, 34, 126, 144, 166, 172, 176, 185, 193, 209
    Nash solution, 34
    set, 3
Baron, D., 49
Baxter Healthcare, 111
Bayes' Rule, 24
Bayesian learning, 205
Bayesian reasoning, 206
BBC, 130
behavior, 76, 127, 171, 187
    strategy, 77
belief, 16, 22, 42, 44, 52, 55
    common, 23
belief revision, 15, 18, 23
Bell Laboratories, 1, 122
Benoit, J.-P., 146
Benthem, J.v., 9, 147
Berg, J., 53
Bernasconi, M., 53
best response, 55
Bethesda, 2
Bhatt, M., 52
biblical studies, 159
Bicchieri, C., 21, 189
Billera, L., 110
Binmore, K., 23, 28, 33, 50, 189, 198, 205, 211
biology, 27, 61, 63, 66, 68, 89, 97, 98, 101, 147, 178, 196, 198, 206, 207, 211
    evolutionary, 111, 116, 120
    population, 75
bisimulation, 16
Bixby, R., 110

Blass, A., 13
Bliss, C., 62
blood-plasma, 112
bluffing, 33
Blume, L., 42
board interlock, 56
Bob, 129
Bonanno, G., 11
Borel, E., 141
botany, 159
Bowles, S., 196
Boyd, R., 196
Brams, S., 147
Brandenburger, A., 41, 205
Brewer, P.J., 170
Brick, A., 65
broadcasting, 87
Brockmann, H.J., 67, 76
Brouwer, L.E.J., 13
Bruin, B.d., 12
Bshari, R., 84
Buchanan, J., 192
Bulfin, R.L., 171
Burke, M.A., 171
Bush administration, 151
business executive, 111

California Institute of Technology, 57
Camerer, C.F., 49, 53, 200
category theory, 9
cell anaemia, 81
"cheaptalk", 54
cheating, 86
chemistry, 102
Cherbourg, 33
chess, 118
Chicago, 113
choice, 129
choice theory, 122
Chong, K., 52
Chopra, S., 147
Church, A., 14

City University of New York, 152, 203
coalitional theory, 5
cognition, 58, 76, 89
cognitive hierarchy, 52, 55
cognitive science, 19
cohesiveness, 54
combinatorics, 97
commitment, 126, 176, 187
communication, 64, 89, 107, 149
competition, 34, 45, 49, 138, 147
  perfect, 3
  price, 3
competitive position, 45
completeness, 44
complexity, 117
computational cost, 6
computer science, 6, 12, 13, 18, 27, 97, 98, 101, 106, 111, 116, 120, 140, 153, 159, 178, 206, 211
Concorde fallacy, 77
Condorcet paradox, 143
confirmation theory, 21
conflict, 78, 89, 125–127, 186
conflict resolution, 78
constitutional principle, 193
contest, 64
contract, 86, 187
contractarianism, 192
convention, 21, 27, 189, 194
  'owner wins', 77
conversation, 149, 160
convexity, 157
cooperation, 64, 89, 125, 127, 147, 186, 197
coordination, 126, 169, 185, 187
core convergence, 46

core stability, 138
Cornell University, 109, 146
Costa-Gomes, M., 52
counterfactual, 24
Crawford, V., 50, 52
criminal coercion, 186
cryptography, 153
'Cultural Park', 10
CUNY Graduate Center, 146, 152

D-structure, 148
Darwin, C., 64, 84
Davies, N., 76
Davis, J., 56
Dawkins, R., 62, 76, 88
Debreu, G., 122, 138
Dechesne, F., 12
decidability, 14
decision theorist, 50, 166
decision theory, 6, 21, 22, 99, 118
  Bayesian, 38, 166
  non-Bayesian, 115
decision-making, 21
Dekel, E., 42, 167
Dekker, P., 11
deliberation, 189
desert spider, 78
design, 64
deterrence, 126
Devetag, G., 57
Dickhaut, J., 53
diffusion of innovation, 56
distributed learning, 211
distributed system, 207
Ditmarsch, H.v., 12
Dixit, A., 102
DNA, 78
dominance, 43
Dubey, P., 110
Dugatkin, L.A., 64
duopoly, 4

Düsing, K., 85
Dutch language, 9
Dvoretzky, A., 157
dynamic model, 98
dynamic programming, 122
dynamics, 106, 209

*Econometrica*, 101
economics, 3, 27, 34, 50, 62, 90, 98, 110, 111, 116, 125, 137, 152, 159, 165, 174, 187, 190, 191, 196, 198, 206, 211
    behavioral, 4, 104, 149, 191
    behavioral , 89
    classical, 4
    empirical, 5
    experimental, 4, 82, 191, 200
    micro-, 165
    normative, 191
    philosophy of, 191
    welfare, 191
economist, 33, 34, 56, 187
economy, 138
    exchange, 138
    political, 49
education, 83
effector mechanism, 88
egalitarianism, 36
Ehrenfeucht, A., 11, 13, 145
election, 4, 151
electricity, 4
elimination of option, 187
Elster, J., 208
empirical reality, 199
employment, 124
'endowment effect', 104
engineer, 119
English Channel, 33
epidemiologist, 56

epistasis, 80
epistemology, 22, 37, 205
    interactive, 2, 19, 22, 101, 111, 115
equilibrium, 123, 167, 175, 205, 209
    Bayesian Nash, 205
    behavior, 123
    competetive, 2, 86, 170
    correlated, 3, 83, 129
    dual, 186
    efficient, 58
    epistemic foundations of, 129
    Markov, 174
    mixed, 43, 169
    Nash, 3, 25, 34, 36, 43, 55, 57, 78, 83, 98, 117, 149, 203
    preference-based, 18
    price, 6
    pure, 43, 169
    quantal response, 55, 174
    "state of mind", 52
    strategic, 14
    subgame-perfect, 34
    tools, 142
equity, 54
equivalence result, 46
equivalence theorem, 2
Eshel, I., 82
ESS, *see* game theory, evolutionary
ethics, 27, 35, 194
    naturalized, 28
Euclid, 10
European Science Foundation, 12
evolution, 36, 80, 106, 197
    adaptive phenotypic, 81
evolutionary dynamics, 99
evolutionary model, 30

expectation, 186
experience, 196
experimentation, 50, 51
extinction, 64

fair division, 139, 147
fairness, 109, 144, 204, 208, 211
fallacy of composition, 186
FDA, 113
Fehr, E., 196
Feiler, L., 52
Feldman, M.W., 82
fighting ability, 79
firm, 123, 175
Fisher, R.A., 64, 85
fitness, 64, 66, 79, 80
fixed-point theory, 18
fMRI brain imaging, 52
focal point, 127, 186, 195, 199
Folk Theorem, 2, 35, 123, 127, 132, 169
formalization, 64, 102
Forrestal Research Center, 1
forward induction, 26
Foster, D.P., 205, 209
Friedenberg, A., 42
Friedman, J.W., 171
Fudenberg, D., 49, 123, 127

Gale-Shapley algorithm, 139, 144, 203
game, 145, 160
    Abelard-Eloise, 145
    argumentation, 10
    assurance, 57
    battle-of-the-sexes, 172
    Bayesian, 29
    chicken, 87, 126
    combinatorial, 115
    cooperative, 29, 45, 149
    coordination, 41, 169, 172
    dictator, 28
    dynamic, 175
    economic, 115
    empirical, 115
    evaluation, 13
    experimental, 115
    form, 166, 192
    foundations of, 69
    hawk-dove, 64, 83, 126, 172
    knowledge-dependent, 23
    language, 153
    matching, 52, 205
    matching pennies, 172
    matrix, 9
    multi-player, 13
    nematode, 89
    neurological study of, 115
    non-transferable utility, 3
    of imperfect information, 23, 58, 106
    of perfect information, 26
    one-shot, 151, 209
    political, 115
    repeated, 54, 88, 175
    sequential assessment, 71
    social dilemma, 28
    stag hunt, 57
    symmetric, 78
    team, 52
    trust, 28
    two-person zero-sum, 141
    ultimatun, 28
    war of attrition, 172
    zero-sum, 45
game theory, 1, 9, 11, 12, 14, 21, 33, 34, 36, 38, 50, 53, 55, 69, 75, 122, 131, 157–159, 165, 166, 185, 187, 189
    algorithmic, 106, 115
    analytical application, 175

analytical implication, 171
as model, 27
as umbrella, 4, 103
behavioral, 115, 161
biological, 61, 66
classical view, 166
conceptual application, 177
conceptual implication, 172
cooperative, 6, 41, 45, 86, 100, 109, 115, 143, 204, 207
eductive, 37
epistemic, 42, 46
evolutionary, 19, 28, 37, 53, 75, 87, 115, 166, 171, 189, 195, 209
experimental, 28, 30, 162
foundations of, 37, 42, 52, 111, 129
"humanistic view", 167
interpretation of, 161
methodology of, 200
neuro, 162
non-cooperative, 6, 27, 43, 100, 115, 209
normative implication, 170
positive application, 173
positive implication, 171
real-life application, 110
theoretical application, 110
unified field theory, 3
Game Theory Society, 115
"gameist", 187
Gärdenfors, P., 15, 24
Gardner, M., 61
Geanakoplos, J.D., 45, 145
generalizability, 51
genetics, 66, 76, 80, 82
geometry, 109
German Fraunhofer Foundation, 11
Gibbard, A., 204

gift exchange, 58
Giles, R., 11
Gintis, H., 196
Girard, J.-Y., 13
Glazer, K., 160
Gluzman, M., 147
Gödel, K., 14
Goeree, J.K., 174
government, 35
GPS, 107
Grafen, A., 61, 76
gravitation, 4
Gray, J., 44
Greek Antiquity, 10
Greenwald, A., 146
Grice, P., 148
group selection, 197
Gul, F., 167
Gurven, M., 197

*Haaretz*, 158
habitat selection, 64
Hagen, E., 86
Hahn, F., 41, 122
Haile, P.A., 171, 174
Halevy, Y., 178
Halpern, J., 11
Hamilton, B., 61
Hammerstein, P., 66, 75, 171
handicap, 64
handicap principle, 65, 83, 84
Harrenstein, P., 12
Harrington, J.E., 171
Harsanyi value, 3
Harsanyi, J., 36, 43, 100, 105, 122, 132, 134, 141, 167, 169, 193, 209
Hart, S., 3, 97
*Handbook of Game Theory*, 3, 103
Hashimoto, T., 204
Hay, D., 61
Healy, P., 58

Heath, A., 61
Hebrew University, 2, 157
Hindu theology, 13
Hintikka, J., 11, 13, 145
history, 159
HMO, 112
Ho, T., 52, 55
'Hochschultaschenbücher', 10
Hodges, W., 13
Hoek, W.v.d., 11
Holmstrom, B., 49
Holt, C.A., 174
Homo Economicus, 87, 149
Homo Ludens, 87
Hortacsu, A., 174
hostage, 187
Hume, D., 194, 211

IBM, 146
ILLC, 11
imitation, 30
Immelmann, K., 75
immune system, 88
immunology, 76
implicature, 148
IMSSS, 122
industrial organization, 138
inequity-aversion, 54
inference, 159
information, 16, 24, 29, 65, 70, 77, 106, 128, 134, 148, 171
    change, 16
    imperfect, 11, 14
    incomplete, 171, 175
    perfect, 175
    transfer, 150
inquiry, 13
interaction, 12, 78, 128, 168, 172, 186, 188, 193, 210, 211
    human, 10
    intelligent, 12, 19

social, 53
strategic, 57, 116, 173, 177, 198
invariance, 13
IO, 49
Iran, 158
irrationality, 167
Israel, 2, 110, 158

Jackson, M., 56
Jerusalem, 102, 157
job referral, 56
Johnson-Laird, P., 57
Johnstone, R.A., 71
Jonker, L.B., 66
Jordan, J.S., 205
justice, 27
justification, 22

Kahneman, D., 4, 104, 149, 188
Kalai, E., 49, 109, 205
Kalai-Smorodinsky solution, 208
Kamien, M.I., 171
Kandori, M., 209
Kaplan, H., 197
Karlin, S., 80
Keisler, H.J., 42
Kelly, K., 15
Kemeny, J.G., 9
    *Finite Mathematics*, 9
Kiefer, J., 109
Knez, M., 57
knowledge, 16, 22, 44, 111, 150
    common, 21, 23, 30, 37, 50, 129, 195, 200
Kohlberg, E., 43, 97, 167
Kosenok, G., 174
Kozen, D., 146
Krasucki, P., 146
Krebs, J., 76

Kreps, D.M., 121, 172
Kuhn, T., 21
Kurz, M., 122

language, 36, 140
   mathematical, 54
   natural, 11
Latin, 83
law, 187, 188
learning, 30, 52, 57, 64, 87, 140
   boundedly-rational, 200
learning dynamics, 190
learning theory, 15
Ledyard, J.O., 170
legislative apportionment, 203
Lehrer, E., 205
Lensberg, T., 209
Levi, I., 24
Lewis, D., 15, 28, 195, 205
Lewontin, R., 63
lexicographic probability system, 43
Liar paradox, 61
Liberia, 177
liberty, 192
limited memory, 87
linguistics, 19, 148, 160, 198
List, J., 104
logic, 9, 12, 97, 101, 109, 145, 159, 187
   and notions of control, 14
   computational, 16
   default, 25
   dynamic, 146
   dynamic epistemic, 18
   epistemic, 11
   finite information, 148
   first-order, 13, 24, 145, 147
   fixed-point, 18
   IF, 145
   intuitionistic, 148
   mathematical, 16
   modal, 19
   monotonic, 24
   philosophical, 16
   philosophy of, 13
   propositional dynamic, 146
   *-, 148
logical positivism, 62
logical validity, 10
'LogiCCC', 12
logician, 9, 187
London, 185
London School of Economics, 33
Lorenzen, P., 10, 13
   *Logische Propädeutik*, 10
Lucas, W., 109
Luce, R.D., 2, 185, 203
   *Games and Decisions*, 10

Machiavelli, N., 37
magnetism, 4
Mailath, G.J., 169, 175, 176, 209
Malmö, 204
Malthus, T., 84
management, 159
Mannheim, 10
'Many Body Problems', 13
'Many Mind Problems', 13
marginal contributions principle, 204
market, 84
   biological, 84
   competitive, 165
   credit, 86
   free-form, 45
   labor, 85
   large, 3
   matching, 6
   oligopolistic, 3
market model, 45

market structure, 45
Mars, 86
Martian, 87
Mas-Colell, A., 98
Maschler, M., 127, 207, 209
Maskin, E., 127
matching socks, 152
mate choice, 56
mathematics, 9, 13, 42, 44, 82, 97, 101, 103, 110, 111, 137, 163, 178, 187, 211
    applied, 55, 140
    foundations of, 16
Mathews, S.A., 171
maximand, 67
maximization, 168, 192
Maynard Smith, J., 37
McCabe, K., 53
McKelvey, R.D., 174
mechanism, 197
    design, 35, 58, 208
medicine, 89, 159
mental representation, 57
Mertens, J.-F., 43, 167
methodological individualism, 140
methodology, 4, 18, 102
Meulen, A.t., 11
Milgrom, P., 49, 122, 171, 172
Milinski, M., 76
military strategist, 188
milk producing gland, 78
minimax theorem, 137
Mirman, L.J., 171
Mirrlees, J., 62
Mishra, B., 146
MIT, 1
*MIT Technology Review*, 106
model, 118, 168, 177
model theory, 10
modelling, 137

Moebius strip, 61
monopoly, 3, 49, 138
    bilateral, 45
monotonicity, 204
moral rule, 27
Moran, P.A.P., 80
Morgenstern, O., 3, 33, 41, 45, 165
morphology, 75
Morris, M., 56
Moulin, H., 137
music, 89
Myerson, R., 34, 49

Nachbar, J.H., 205
Nagel, R., 52, 55
Nash bargaining solution, 208–210
Nash, J., 1, 34, 55, 100, 132, 134, 165, 209
National Bureau of Standards, 2
natural science, 165
natural selection, 63, 70, 80, 87, 99, 196
natural selection theorem, 64
naturalistic fallacy, 28
network, 56, 89
    formation, 56
Neumann, J.v., 33, 41, 45, 141, 165
    *Theory of Games and Economic Behavior*, 3
Neumann-Morgenstern stable set, 3
New Mexico, 79
New York University, 146
*Newsweek*, 35
Newton, I., 118
Nobel Laureates in Economics, 121
Nobel Prize Committee, 132
Nobel Prize Foundation, 129

Nobel Prize in Economics 2005, 4, 125
Noë, R., 84
norm, 54, 194, 209
  emergence of, 211
  fairness, 36
Northwestern University, 49, 110
NP-complete, 152
NSF, 159
nuclear weapon, 1
nuclear-weapons policy, 186
nucleolus, 98
number theory, 153

Okada, N., 204
oligopoly, 3, 45
omniscience, 30
open-air market, 157
operations research, 1, 111, 122, 140
optimization, 7, 63, 66, 87, 101, 113
organization, 123, 177, 208
organized crime, 186
Osborne, M.J., 146
Oslo Peace Accords, 132
Otterloo, S.v., 12
Owen, G., 110, 137
owner-intruder asymmetry, 83
ownership, 77

Pacuit, E., 147
Palfrey, T.R., 174
parasitic nematode worm, 88
parent-offspring conflict, 64
Pareto optimality, 86, 144
Parida, L., 152
Parikh, P., 149
Parikh, R., 15, 30, 145
Parker, G., 76
Pathak, P., 170
Pauly, M., 12, 147

Peirce, C.S., 148
Peleg, B., 157
Peranson, E., 171
Perry, M., 99
persuasion, 160
phenotypic plasticity, 89
philosophy, 9, 19, 22, 27, 101, 111, 178, 206, 211
  of language, 28, 197
  of science, 11, 21
  political, 28
physicist, 119, 187
physics, 187
physiology, 75, 77
Plame, V., 151
Plato, 11
pleiotropic gene effect, 80
Plott, C.R., 170
Poker, 33
Polemarchakis, H.M., 145
political science, 35, 98, 101, 110, 111, 116, 125, 141, 159, 178, 190, 211
Porter, D., 170
Postlewaite, A., 46
pragmatics, 148, 160
Pratt, V., 146, 152
predator-prey, 64
prediction, 78
preference
  change, 16
  individual, 191
  social, 54, 191
Premack, D., 46
Price, G., 61, 75
Princeton University, 1
prisoners' dilemma, 29, 54, 61, 87, 118, 122, 128, 151, 159, 172, 177, 186, 198
probability theory, 42, 97, 116

process structure, 15
Professor Moriarty, 46
prominence, 199
proof, 10, 13
   deductive, 10
   formal, 146
proof theory, 10
psychology, 50, 58, 62, 82, 103, 120, 125, 153, 187, 198, 200
   social, 54, 124
punishment, 90

quantum mechanics, 15

R.D. Luce, 49
Rabin, M., 157
racial segregation, 186
Radner, R., 122
Raiffa, H., 2, 49, 185, 203
   *Games and Decisions*, 10
Ramsey, F., 148
RAND, 55, 186
randomization, 43
Rassenti, S.J., 171
rational choice, 187
rationality, 21, 22, 37, 42, 55, 58, 70, 82, 86, 104, 116, 141, 145, 159, 192, 193, 211
   act, 5
   Bayesian, 205
   bounded, 117, 162
   common assumption of, 44
   perfect, 166, 195, 200
   rule, 5
Rawls, J., 36, 193
reasonableness, 209
"reciprocal fear of surprise attack", 186
reciprocal promise, 187
reciprocity, 54, 89, 196

recombination, 80
Reeve, H.K., 64
Reny, P., 23
Renyí, A., 43
replicator dynamics, 30, 53, 87
representation, 175
reputation, 123, 134, 175
Rick, S., 52
Riechert, S., 65, 78, 171
Rob, R., 209
Roberts, J., 49, 122, 171, 172
Rooij, R.v., 15
Rosenthal, R.W., 23, 46
Roth, A.E., 6, 170, 171
Rothschild, M., 122
Royal Swedish Academy, 125
Rubinstein, A., 5, 34, 50, 127, 145, 146, 157, 166, 171, 176, 210

Salame, S., 147
Samet, D., 114
Samuelson, L., 50, 57, 169, 174–176
Sandu, G., 11, 145
Santa Fe bar problem, 146
Satterthwaite, M.A., 204
Savage, L., 22, 38, 99, 146
*Scandinavian Journal of Economics*, 102
Scarf, H., 138
Schelling, T.C., 4, 102, 121, 125, 134, 137, 176, 185, 193
Schmeidler, D., 97, 110, 207
Schulte, O., 25
Schwartz, N.L., 171
Seger, J., 67
selection, 90
self control, 186
self-interest, 127
selfish genetic element, 80

Selten, R., 67, 76, 87, 132, 134, 141, 167, 169, 209
semantics, 145, 148
  game-theoretical, 11
Sen, A., 62, 191, 193
Sevenster, M., 12
Shaked, A., 50
Shapley value, 3, 6, 114, 139, 203, 208
Shapley, L., 110, 127, 204, 207
Sherlock Holmes, 46
Sheshinski, E., 122
sibling rivalry, 64
signalling theory, 83
Simon, H., 124
Skolem function, 15
Skyrms, B., 28, 189
Slovic, P., 149
Smith, A., 195
Smith, J.M., 61, 75, 101, 195, 207
Smith, V.L., 4, 171
Snell, J.L., 9
Sobolev, A.I., 209
social choice, 111
social choice theory, 191
social contract, 27, 193
social contract theory, 189, 193
social engineering, 144
social foraging, 64
social image, 54
social learning, 87
social norm, 21, 28, 189
social science, 27, 103, 116, 137, 165, 206
social scientist, 187, 188
social software, 19, 152
sociological theorist, 187
sociologist, 54, 56, 187
sociology, 124, 125, 187, 190
solidarity, 54

solipsism, 13
Sönmez, T., 170, 171
Sophism, 11
Sorites paradox, 149
Sotomayor, M., 6
specialization, 119
Spence, M., 83, 122
Spinoza Award Project, 11
St Francis of Assissi, 38
Stahl, D., 55
Stahl, I., 210
Stalnaker, R., 11
Stanford University, 12, 122
statistical physicist, 56
statistical theorist, 187
statistician, 187
statistics, 75, 109, 116, 140, 187
Stearns. R., 127
Stiglitz, J., 62, 85, 122
strategic awareness, 64
strategic reasoning, 22
strategy, 1, 13, 18, 26, 36, 42, 77, 83, 148, 159, 196
  business, 44
  conditionality, 77
  evolutionary stable, 61, 69, 77, 80
  nuclear defense, 186
  of learning, 113
  optimal, 137
  parasite, 89
  tree, 112
strategy-proofness, 139, 144
'streetcar theory', 66, 81
Strotz, R.H., 128
Stuart, H., 44
subgroup consistency principle, 209
Sugden, R., 28, 191
sure thing principle, 99, 146
survival probability, 79

Sutton, J., 50

Tamer, E., 171
TARK, 11
Taylor, P.D., 66
Tel Aviv University, 97, 110
Telser, L., 49
territoriality, 78
"the impossibility of a Paretian liberal", 191
'the winner's curse', 87
theoretical physicist, 187
"theorists", 187
"theory", 187
theory of mind, 46
Thompson, G.L., 9
threat, 187
Tirole, J., 49
topology, 109
Torres, F.o.d., 67
trade, 4, 84, 105, 138
    international, 3
transaction, 86
transportation, 107
tremble, 210
trust, 53
truth, 13, 22, 159
Turing machine, 14
Turing, A., 14
Tversky, A., 104, 149, 157

Ullman-Margalit, E., 28
UN, 6
uncertainty, 116, 123, 129, 193, 205
unilateral promise, 187
University of Amsterdam, 12
University of Bielefeld, 67, 75
University of California, Berkeley, 122
University of Cambridge, 21, 41
University of Chicago, 49

University of Oxford, 61
Unver, U., 171
US Congress, 6
US Defense Department, 1
USA, 177, 186
USSR, 177, 186
utilitarianism, 36

Väänänen, J., 13
Vanderschraaf, P., 28
"veil of ignorance", 193
Venema, Y., 11
Vilks, A., 11
Virginia Polytechnic Institute, 192
von Neumann, J., 211
voting, 151, 203
voting theory, 147

Walrasian analysis, 86, 141
Walzer, M., 208
Warglien, M., 57
Washington, 2
Wassenaar, 150
Web agent, 29
Weber, R., 52, 110
Weigelt, K., 53
welfare judgment, 193
Wessen, R., 170
West-Eberhard, M.J., 75
Wilson, R.J., 122, 172
Wittgenstein, L., 153
Wolinsky, A., 56
Woodruff, G., 46

Yaari, M., 157
Yale University, 185
Young, H.P., 68, 152, 171, 189, 203

Zahavi, A., 65, 83
Zajac, E.E., 208
Zeckhauser, D., 102

Zermelo, E., 26
zoology, 62, 159

www.ingramcontent.com/pod-product-compliance
Lightning Source LLC
Chambersburg PA
CBHW020752160426
43192CB00006B/308